BLOCKBUSTER SCIENCE

THE REAL SCIENCE IN SCIENCE FICTION

DAVID SIEGEL BERNSTEIN

59 John Glenn Drive
Amherst, New York 14228

Published 2017 by Prometheus Books

Blockbuster Science: The Real Science in Science Fiction. Copyright © 2017 by David Siegel Bernstein. All rights reserved. No part of this publication may be reproduced, stored in a retrieval system, or transmitted in any form or by any means, digital, electronic, mechanical, photocopying, recording, or otherwise, or conveyed via the Internet or a website without prior written permission of the publisher, except in the case of brief quotations embodied in critical articles and reviews.

Trademarked names appear throughout this book. Prometheus Books recognizes all registered trademarks, trademarks, and service marks mentioned in the text.

Cover design by Jacqueline Nasso Cooke
Cover images courtesy of NASA and © Shutterstock
Cover design © Prometheus Books

Inquiries should be addressed to
Prometheus Books
59 John Glenn Drive
Amherst, New York 14228
VOICE: 716–691–0133
FAX: 716–691–0137
WWW.PROMETHEUSBOOKS.COM

21 20 19 18 17   5 4 3 2 1

Library of Congress Cataloging-in-Publication Data

Names: Bernstein, David Siegel, 1967- author.
Title: Blockbuster science : the real science in science fiction / David Siegel
    Bernstein.
Description: Amherst, New York : Prometheus Books, 2017. | Includes index.
Identifiers: LCCN 2017021982 (print) | LCCN 2017035332 (ebook) |
    ISBN 9781633883703 (ebook) | ISBN 9781633883697 (hardback)
Subjects: LCSH: Science fiction—History and criticism. | Science in literature.
    | Science fiction films—History and criticism. | Science in motion pictures.
    | Science fiction television programs—History and criticism. | Science in
    popular culture. | BISAC: SCIENCE / Philosophy & Social Aspects. | FICTION
    / Science Fiction / General.
Classification: LCC PN3433.6 (ebook) | LCC PN3433.6 .B466 2017 (print) |
    DDC 809.3/8762—dc23
LC record available at https://lccn.loc.gov/2017021982

Printed in the United States of America

For my mother, Norma, who infected me with the writing bug, and my sister, Sarah, who believed in me more than anyone.

CONTENTS

Introduction	17

CHAPTER 1. ONCE UPON A SPACETIME 27

Einstein Considers the Gravity of His Ideas	30
Mass Is Energy, and Sometimes It Makes Holes	32
No Witnesses . . . Yet	33
Two Scientifically Plausible Ways to Surf Spacetime Waves	34
Parting Comments	37
Chapter 1 Bonus Materials	37
Bonus 1: A Special Relativity Paradox	37
Bonus 2: How Much Energy Is Contained within the Matter of Your Body?	38
Bonus 3: Mutants	39

CHAPTER 2. IF YOU ARE UNCERTAIN, CALL A QUANTUM MECHANIC FOR A FIX 41

What Is a Quantum Leap?	44
That Special Moment . . . When You Just Know	44
Real Science and Science Fiction	46
Mind-Blowing Quantum Trivia	48
Parting Comments	48
Chapter 2 Bonus Materials	49
Bonus 1: A Time-Travel Paradox	49
Bonus 2: Photosynthesis and Quantum Mechanics	50
Bonus 3: What Are Quantum Suicide and Quantum Immortality?	50
Bonus 4: Time-Traveling Text Messages	51
Bonus 5: Subatomic Uncertainty in the Classical World	51

FIRST INTERLUDE: A TOUCH OF ATOMIC THEORY — 53

- The Four Universal Forces — 54
- What Is an Atom? — 55
- How Old Is the Oldest Atom? — 56
- What's the Matter with Antimatter? — 56
- Going Deep — 57
- First Interlude Bonus Materials — 59
  - Bonus 1: The Range of Influence of the Four Universal Forces — 59
  - Bonus 2: Nuclear Fusion versus Nuclear Fission — 59

CHAPTER 3. STRUMMING OUR WAY INTO EXISTENCE — 61

- Extra Dimensions — 62
- Meet the Competition — 64
- Can Quantized Space Solve a Paradox and Hurt a Villain? — 65
- Parting Comments — 66
- For the Record — 66

CHAPTER 4. OUR UNIVERSE (AS OPPOSED TO THOSE OTHERS) — 67

- How Big Is the Universe? — 67
- How Is the Observable Universe Different from the Actual Universe? — 69
- How Are Galactic Distances Calculated? — 70
- And in the (or Rather a) Beginning . . . — 71
- What Caused the Big Bang? — 73
- What Happened Next? — 75
- What Are Gravitational Waves? — 76
- Have Gravitational Waves Ever Been Detected? — 77
- Some General Facts about Our Universe (as We Know It) — 77
- Our Solar System, Home of Sol, Our Sun — 79

Parting Comments	80
Chapter 4 Bonus Materials	81
Bonus 1: Olbers's Paradox	81
Bonus 2: Your Suntan	81
Bonus 3: The Brief History of Our Sun	81
Bonus 4: The On and Off Lighting System of a Cepheid Star	82

CHAPTER 5. PARALLEL WORLDS 83

Parallel Worlds from Math	84
Parallel Worlds from Distance	84
Parallel Worlds from Branching	85
Mem(brane) Theory	86
The *We Live in the Best of All Worlds* Theory	87
Parting Comments	88

CHAPTER 6. POWERING UP OUR CIVILIZATIONS 91

Civilization Rankings	92
Are There Quantum Energy Sources?	94
Is There Evidence of Virtual Particles?	95
Virtual Particles and Zero-Point Energy in Science Fiction	95
Parting Comments	96
Chapter 6 Bonus Materials	97
Bonus 1: An Inward Look: An Alternative Classification of Civilizations	97
Bonus 2: Comparing Energy Sources	97

CHAPTER 7. BLACK HOLES SUCK 99

How Do Black Holes Arise?	99
Why Does Escape Velocity Matter?	102
How Do Black Holes Relate to Spaghetti?	103
Do Black Holes Last Forever?	103

How Are Black Holes Detected?	104
Wormholes (Again)	105
Can Black Holes Be Used for Time Travel?	105
Parting Comments	107
Chapter 7 Bonus Materials	108
Bonus 1: A Massive Mystery	108
Bonus 2: The Information Paradox	108

CHAPTER 8. ORIGIN AND EVOLUTION OF LIFE ON EARTH   109

The Human Story	111
Fossils Are So Yesterday (Literally)	114
Whoa! Slow Down and Keep It Real	114
How Did Life Begin on Earth?	115
What Is DNA?	116
Parting Comments	120
Chapter 8 Bonus Materials	120
Bonus 1: Evolution Is Smart	120
Bonus 2: The "Ome" Home	121

SECOND INTERLUDE: MORE OR LESS HUMAN   123

CHAPTER 9. BADASS BIOLOGY   125

Genetic Engineering and Evolution	125
How Is Genetic Modification Done?	126
What Does Science Fiction Say about the Ethics of Gene Modification?	126
Speaking of Immortality, Is It Possible?	127
Is a Longer Life Worth It?	129
Can't Get Enough of Yourself? Send in the Clones	130
Genetics for the Zombie Apocalypse	131
Bacteria and Viruses to the Rescue	132
Parting Comments	133

Chapter 9 Bonus Materials	133
Bonus 1: Crime-Busting Science	133
Bonus 2: Metamorphosis	134

CHAPTER 10. WELCOME TO TECH U 135

Celebrity Fictional Cyborgs	137
The Transhuman Brain	138
Sometimes You Need to Go Small	139
And in the End, Posthumanism	140
Cybering (Everyone Is Doing It)	141
Parting Comments	142
Chapter 10 Bonus Materials	142
Bonus: Nano for Food	142

CHAPTER 11. MAN AND NATURE 145

Can Climate or Weather Be Predicted?	146
What Is the Greenhouse Effect?	147
People Who Live in Glass Houses Shouldn't Throw Carbon	148
How Is the Average Temperature of the Earth Calculated?	149
Is There Evidence of Global Warming?	150
Is Anything Being Done to Prevent Global Warming?	151
Let There Be (a Little) Light	152
What Other Human Activities Affect Weather?	152
Is Fiction Being Used to Help Warn about Global Warming?	153
Parting Comments	154
Chapter 11 Bonus Materials	155
Bonus 1: Negative Greenhouse Effect	155
Bonus 2: A Dawning of a New Age (Really, an Epoch)	155
Bonus 3: The Ozone Layer	156

CHAPTER 12. TIME TO MOVE (PLAN B) 157

What Basic Planetary Elements Make for Happy Colonists?	159
Extra Resources: Can Asteroids Be Used to Our Advantage?	162

Thinking Galactically but Searching Locally for a
   Vacation Home   163
Searching Galactically: What Is an Exoplanet and
   How Do We Find One?   165
To Invade or Not to Invade?   167
Parting Comments   167
Chapter 12 Bonus Materials   168
   Bonus: How Do We Handle All the Radiation the
     Universe Throws at Us?   168

CHAPTER 13. INTELLIGENCE COMES IN ORGANIC AND ARTIFICIAL FLAVORS   171

What Do Emergent Properties Have to Do with My Thoughts?   172
How Do Neurons Create Consciousness?   173
Where Does Cognition Fit?   174
What Is Intelligence?   175
The Importance of Creativity
   (Something Your Neurons Might Do for Fun)   175
Rise of the AI   176
Should We Put It to the Test?   177
What Is the Technological Singularity?   178
Ethical Concerns of a Post-Singularity World   179
Full Circle   180
Parting Comments   181
Chapter 13 Bonus Materials   181
  Bonus 1: Moore's Law   181
  Bonus 2: Time Is in the Eye of the Beholder   182
  Bonus 3: Nefarious Software
    (Three Annoying Computer Infections)   183

CHAPTER 14. THE RISE OF THE ROBOTS   185

In the Beginning, There Was a Word and a Few Wires   186
Robotic Evolution   187
To Serve and Obey   188
Robots in Our Everyday Lives (as Servants)   191

    Question of Ethics   194
    Parting Comments   194
    Chapter 14 Bonus Materials   195
        Bonus: A Few Celebrity Science Fiction Robots   195

CHAPTER 15. ARE WE ALONE? EXTRATERRESTRIAL INTELLIGENCE   197

    Never Mind Meeting Them. Do Aliens Exist?   199
    So, Why Haven't We Heard Anything from Aliens?   200
    Perhaps Earthlike Planets Are Rare   202
    Any Ideas about What Alien Life Looks Like?   204
    Parting Comments   205
    Chapter 15 Bonus Materials   206
        Bonus 1: Ufology   206
        Bonus 2: Carbon and Water, a Marriage Contract for Life.
            Is It Recognized Everywhere?   206
        Bonus 3: A World Protocol for Extraterrestrial Contact   207

CHAPTER 16. A REALLY LONG-DISTANCE CALL: INTERSTELLAR COMMUNICATION   209

    Theoretical Long-Distance Calls   210
    Parting Comments   216

CHAPTER 17. *AD ASTRA PER ASPERA*: A ROUGH ROAD LEADS TO THE STARS   217

    Keeping It Local: Atmospheric Travel   218
    The Long Journey: Space Travel   219
    Gravity and the Space Traveler   225
    If Science Fiction (Adventure) Uses Artificial Gravity,
        Then Why Not . . .   228
    A Hopefully True Story: A Tale of Application   228
    Parting Comments   229

THIRD INTERLUDE: A MATTER OF SUBSTANCE — 231

    So, What Is Mass? — 232
    Why Do We Have Mass? — 232
    What Makes Dark Matter and Dark Energy So . . . Um . . . Dark? — 233
    Is There Evidence of Dark Matter? — 234
    Sometimes Substances Like to Change Their Outfits Before Going Out — 235

CHAPTER 18. WHY ARE WE SO MATERIALISTIC? — 237

    What Is Material Engineering? — 237
    Are Any New Materials under Development Now? — 238
    Speaking of Athletes, Here Are a Couple of (Literally) Cool Proofs of Concept — 239
    It's Elementary — 240
    What Is a Superconductor? — 241
    Any Way around That Freezing Problem for Superconductors? — 242
    A Carbon Solution for Materials — 242
    Can Any Materials Harvest Light? — 243
    Now You Saw It, and Now You Didn't — 244
    Parting Comments — 246

CHAPTER 19. TECHNOLOGY (COOL TOYS) — 249

    The Laser — 250
    Popular Names for Light-Energy Weapons in Science Fiction — 250
    3-D Printing — 251
    How about a Technology Combo Meal with a 3-D Printer? — 252
    Great Tool, but Are There Ethical Issues with 3-D Printing? — 252
    Wearable Technology — 253
    Camera Technology — 253

Computer Identification	254
Machine Mind Control	254
Parting Comments	255

CHAPTER 20. WHAT DOES IT TAKE TO EXIST? 257

Do Any of Our Ideas about Reality Matter?	258
What Is Augmented Reality?	259
What Is Virtual Reality?	260
When Did the Science of Virtual Reality Begin?	261
Any Practical Uses for VR Unrelated to the Military and Entertainment?	262
Can False Memories Be Considered Virtual Reality?	263
What about Erasing Memories?	264
Holographic Reality	265
Let's Get Deep into Some Science Fiction	266
Holographic Theory and Black Holes	267
Parting Comments	267

CHAPTER 21. THE END OF EVERYTHING 269

Speaking of Endings (an Author's Confession)	269
The End of the Individual	270
The End of the Human Species	270
Roid Rage (Destruction by Asteroid)	272
Is Anyone Watching Out for These Rocks and Ice Balls?	275
If the Bigger NEOs Target Earth, Can They Be Stopped?	275
The Sun and Earth: A Relationship That Ends When the Lights Go Out	277
Why Will Only Supersized Galaxies Exist in the Future?	278
We Came Along at a Good Time (to Predict the End of the Universe)	279
Thermodynamics Will Be the End of Us, but First Learn Its Laws	280
How Does Entropy Explain the Direction of Time?	281
How It All Ends	281

When All the Suns Turn Out the Lights	282
Life after This Universe	282
Parting Comments	283

Acknowledgments	285
Notes	287
Glossary	309
Reading/Movie/Song List	317
Index	331

INTRODUCTION

Everything starts as somebody's daydream.
   —Larry Niven, science fiction author who has big dreams

Science was many things, Nadia thought, including a weapon with which to hit other scientists.
   —Kim Stanley Robinson, *Red Mars*

Have you ever dreamed about voyages between worlds in starships fully equipped with faster-than-light warp drives? Have you ever dreamed about battling aliens from other dimensions, or traveling to a parallel universe to defeat your evil twin and rescue the princess or prince? How about a world where princesses and princes can't be differentiated by biology because gender is flexible?

I'm guessing that if you are reading this, then you must have. The good news is that although some of these dreams are unlikely, all are scientifically possible by extrapolating from today's technology.

The power of stories to inspire us has held true for all of human history. Today, science fiction has the ability to inspire breakthroughs that change our world. Companies used words like "robot" and "android" after they were popularized in fiction. Our STEM experts often say they were first inspired by stories they read when they were young.

This book exists to help you to understand a few of the more popular topics in science as well as how they are used (and sometimes misused) in science fiction. This book isn't only for science fiction fans who want to know more about the science behind the plot. This book is for the curious—*anyone* who wants to know more about the natural world and the universe of which they are a part. It's for the science geek in everyone.

Throughout each chapter, you will find a number of question marks. Many recent discoveries have led to questions that scientists never thought to ask before. Curiosity about our world drives fiction authors and filmmakers to explore the realm of possibility. Besides, isn't science itself all about asking questions? Another thing to beware of is that this book contains spoilers. I only trend toward spoilage when it's necessary to make or fully explain a scientific point about something in fiction.

Science fiction is about change, a world (or worlds) yet to be. Science fiction can explore a hopeful world where problems are solved, or a dangerous (dystopian) world where problems are caused, or a world of existential threats such as drastic climate change or destruction by asteroid (chapters 11 and 21 might cause you some anxiety on these last two topics).

To paraphrase Mark Watney, the intrepid engineer-agriculturist of Andy Weir's novel *The Martian*: in the face of overwhelming odds, humans have scienced the shit out of a lot of problems. This book is packed with examples.

What if change is impossible, unlikely, or even unimaginable? By today's standards, it's difficult to imagine a time when the idea of a different type of future (socially, technologically, politically) would be alien. This is not a fantasy, however. It's human history.

There was a time when the things that our ancestors observed just didn't change all that much. The idea of wildly different futures didn't exist in the age of limited science. Plenty of fiction lurked about, including the magical thinking of fantasy, but stories about radically different tomorrows due to changes in technology, not so much. Back then, people usually didn't travel far from their towns or villages. Mostly they performed the same daily tasks that their parents and grandparents had, and they performed those tasks in the same way.

Enter the Industrial Revolution. The technology and social norms that arose during this period were radically different than they had been in the previous century. After this time, extreme lifestyle changes that occurred within a single generation became commonplace. During this new age, the common wisdom among the common people was that the future was going to be different. But different in what ways?

Welcome, science fiction. The entire category exists partially thanks to the eruption of Mount Tambora in 1815. This event led to the volcanic winter of 1816, the so-called Year Without a Summer. Living through this year inspired one person to find a way to express a radical future.

Mary Wollstonecraft Godwin, her boyfriend Percy Bysshe Shelley, Lord Byron, Dr. John William Polidori, and Claire Clairmont were all hanging out at a villa in Switzerland one of those chilly summer days. Because it was too cold to go outside, the friends challenged each other to write a ghost story (a magical thinking type of story).

Mary struggled to come up with an idea. Then one night as they sat around talking about the nature of life, a great concept hit her. On that night, Frankenstein's monster was born. Mary and Percy Shelley were married later that year. Mary Shelley completed her novel *Frankenstein; or, The Modern Prometheus* in 1818. The book is possibly the first science fiction novel ever.

By the way, the term "science fiction" made its public debut in 1851 in William Wilson's book of essays, *A Little Earnest Book upon a Great Old Subject.* He writes,

> We hope it will not be long before we may have other works of Science-Fiction [like Richard Henry Horne's "The Poor Artist"], as we believe such books likely to fulfil a good purpose, and create an interest, where, unhappily, science alone might fail.[1]

I like Wilson's thinking. I like it so much that I wrote the book you're holding now.

Anyway, before the end of the nineteenth century, Mary Shelley's science fiction spark grew into a blazing fire.[2] Jules Verne teased people with his extrapolations on technology. He was all about plausibility. Get this: he believed that someday people would zip across the seas in electric submarines. Crazy idea. Right? That was in *Twenty Thousand Leagues under the Sea*, written back in 1870. How about his idea of using solar sails (described in chapter 17) for space travel in his 1865 novel *From the Earth to the Moon*?

Then H. G. Wells arrived on the scene with science fiction that contained a more sociological bent. Like Verne, he wrote about changes in technology, but he was less interested in the plausibility of the science than in how the changes might affect people. These two styles have served as the foundations of science fiction ever since.

Following in the style of Verne's hard (plausible) science, the "Big Three" authors of the early twentieth century emerged. Robert Heinlein took technology that existed, such as the telephone, and made it into something that resembles today's devices, like a mobile phone (*Space Cadet*, 1948). Isaac Asimov went robot crazy starting in 1939 (you will read about robots rising up in chapter 14).

The third author, Arthur C. Clarke, wrote about artificial satellites in a stationary orbit twenty years before a real one orbited Earth. He also had ideas about space elevators (learn more about them in chapter 17). All three were friends, but Asimov and Clarke found it necessary to create the Clarke-Asimov treaty.

Under the terms of the treaty, Asimov was required to insist that Clarke was the best science fiction writer in the world (reserving second-best for himself). Clarke, meanwhile, was required to insist that Asimov was the best science writer in the world (reserving second-best for himself).[3] Which led to the following dedication in Clarke's 1972 novel *Report on Planet Three and Other Speculations*:

> In accordance with the terms of the Clarke-Asimov treaty, the second-best science writer dedicates this book to the second-best science-fiction writer.

So, I'm not the first author writing about science and science fiction who also likes humor. As I mentioned before, Wells demonstrated how not all the science in science fiction needs to be plausible (it just needs to not be mystical). Science concepts in fiction can be a tool with which to explore topics that other types of early fiction often avoided.

Although Clarke's stories are generally known for their hard science, he wasn't shy about tackling social issues. In *Imperial Earth*, released in 1975, the main character is of African descent and his sexuality is flexible. Recall how radical it was back then to provide a starring role for a character who happens to have an ethnic heritage . . . let alone to address the fluidity in human sexuality.

Ursula K. Le Guin is the queen of humanistic science fiction. Her stories explore the cultural and social structures of alien cultures. In her, the heart of an anthropologist blends with the writing chops of a Heinlein. She gives alien cultures consistent motivations, albeit different from the ones most readers have experienced before, and she skillfully makes readers sympathetic to the aliens' cause.

And then there is science fiction that borders on philosophy. Philip K. Dick used fiction as a tool to question identity and what constitutes reality . . . sometimes his own. You might harbor your own doubts about what is real after you read chapter 20, where we question existence. Don't worry. You are real. Maybe.

Okay, I need to take a deep breath. I could continue gushing about great authors, but I'll stop for the sake of keeping this introduction under one thousand pages. I think you get the idea. Science fiction has exploded onto the scene with books and films and television shows based on real science. And it all began with Mary Shelley, a bet, and a cold summer.

I'm thinking that a lot of you might be interested in fantasy. That's cool. Fantasy is a living ancestor of science fiction and graciously stops by several of the upcoming chapters for an occasional visit. Keep in mind that science fiction, unlike fantasy, is about rationality. This book sticks to what can be proven using the scientific method and how scientific concepts and theories have been, or can be, extrapolated for fiction.

Nonetheless, a little magical thinking (ghosts, unicorns, etc.) mixed in with plausible science can be fun. Think about steampunk fiction, where modern technology is realized in Victorian England or the American Wild West. I believe that as long as fantasy follows the internal rules of the imaginary world in which it is set, its family resemblance to science is quite striking. After all, science is a consistent and replicable explanation of the natural world.

In fantasy, the author creates a world with its own set of laws. If these laws are consistently applied, and things that happened to one character must happen to others under the same conditions, then you have the fantasy world's equivalent of science. Let's look at this for an example: On a world called Meryton, a young sorceress (like all the magicians in this world) relies on nonrenewable magic. She is running low because of her wasteful father and must marry well to replenish the family's magical treasure chest. Mayhem ensues. Title: *Pride and Prejudice in Space*. Sorry, I couldn't resist.

Why do I and so many other humanoids love science fiction? And yes, for me, both reading and writing science fiction are a love affair. Here is *my* answer to the question: it is so beloved because of how inclusive it is. Any bookstore (brick-and-mortar or online) will stock authors and characters who represent a full array of racial, ethnic, and gender groups. And because fictional worlds can support a variety of societies and cultures, a multitude of worldviews, norms, and sexual orientations have appeared.

Because of this inclusiveness, the typical science fiction fan isn't the clichéd man-child who lives in his parents' basement. I think the typical science fiction fan is all of us: men, women, engineers, lawyers, scientists, actors, sports stars, and so on. If you are a creator (or aspiring creator), learn who your potential readers might be. Guess what? They can be anyone from anywhere in the world.

Before you plunge into writing a book or reading (or viewing) your next science fiction story, let's separate the science from the fiction. Science asks, "What if?" Fiction speculates on what will happen to people or societies when that thorny question is answered. But what actually constitutes science? Broadly speaking, it's a way of knowing. Scientific methodologies seek the truth of the physical world in which we live. Of course, there are other ways of knowing, such as art (personal truth), religion (revealed truth), and so on, but this book isn't about those. This book is evidence based.

Science has two distinct aspects. First, it is a method, a set of steps used to question phenomena in the natural world through reproducible observations and controlled experiments. The steps are fairly concrete: conduct observation, devise a hypothesis to attempt to explain the observations, experiment in ways that test the hypothesis, formulate a theory, confirm the theory through more experiments, and use the theory for prediction.

But wait, that isn't all. The second aspect of science is that it is also a collective term for the accumulation of what was learned. There is the act of *doing* chemistry, and there is chemistry itself.

Don't worry. This book avoids getting too technical on any particular subject. But, for the strong of heart and the endlessly curious geeks, bonus sections provide extra details and/or further extrapolations on some of the topics. Each chapter also includes examples of the good, the bad, and the ugly use of the chapters' scientific topics in novels, television, and movies. A diverse selection of authors and characters will be heard from through these pages. There is also a glossary at the end of the book listing a few of the noteworthy terms used in the chapters, and a reading/movie/song list of sci-fi-themed books, films, music, etc.

Also, don't worry about the math that occurs here and there, because these references are very limited. Never fear math. It is the language of science. In fact, as with spoken languages, it is fraught with tongue twisters that scientists sometimes take too seriously. On occasion, that has led them astray from the methods and demands of science.

For example, consider string theory (the topic of chapter 3). Intuitively it may seem hard to follow, but mathematically it is internally consistent. It might theoretically explain phenomena that can't otherwise be explained using more conventional models. However, because it isn't observable and because no direct experiments can test the theory, for now it dwells more in the realm of philosophy than science.

For creators of science fiction, though, who cares? Whatever else can be said of string theory, it's a great story generator. I hope you will see from this and other examples how math can nudge science toward the world of science fiction. So remember this mantra: math is good.

Blockbuster Science can be read out of sequence if you want to hone right in on a specific topic, but the chapters do have some tendency to build upon each other. I also encourage speculations—your own, not just those that deal with fiction—throughout the book. I want this to be collaborative.

The first two chapters describe the twin pillars of twentieth-century physics: quantum mechanics and Albert Einstein's theories of relativity. Special and general relativity show us that time is woven into our universe and that our perception of time is correlated with gravity.

Before the theories of relativity were developed, studying the universe was like studying a three-dimensional cube as if it were a square—not too helpful. Einstein united space with time, matter with energy, and everything to gravity. From his theories, scientists have deduced the existence of wormholes, black holes, and the big bang.

Another major topic of relativity is the energy-mass equivalency ($E=mc^2$), which is really amazing, and in that chapter 1, you find out why. Are you interested in time travel and space travel? (Who isn't?) This is the chapter for you (along with a few others that dip into the same topic).

In chapter 2, we learn that size really does matter . . . at least for quantum mechanics. We will consider really small stuff, subatomic small, to discover why uncertainty will always be a part of our universe. This is the area of physics where determinism is dumped. At the tiny scale, everything is fuzzy because tangible particles are also waves.

Wait, it gets weirder. The tangible particle can be anywhere in its own wave, where every occurrence of the particle in its own wave is merely a probability. Until an outcome is observed, all the occurrences exist simultaneously. This fuzziness has challenged the way scientists understand reality (and it might do the same for you). Chapter 2 also considers a couple of the more popular interpretations of what they believe this fuzziness and lack of determinism could mean.

Don't worry about the weirdness. The trickier concepts of quantum mechanics are broken down into bite-sized chunks. By consuming these earlier chapters so daintily, in later chapters you will understand the theory behind topics like zero point energy, virtual particles, quantum entanglement, quantum computing, quantum teleportation, loop quantum gravity, quantum suicide and immortality, and time-traveling text messages.

Topics in later chapters include string theory, parallel worlds, antimatter, neutrinos, tachyons, invisibility, holograms, extraterrestrial life, interstellar communication, bioengineering, terraforming, global warming, cosmology, evolution, the origin of life (carbon-based life, at any rate), rocketry, genetic modification, thermodynamics, the "arrow of time," what might be next for computers (artificial intelligence), ranking civilizations, plus much more.

Geez, that's a lot. We are going to have fun! At least until the last chapter, which covers the end of everything—the earth, the sun, the universe . . . everything.

I hope this book will give you a sense of wonder and the desire to explore these topics even more.

The moral of this introduction: Science tells us what is, not what we want. Science fiction has no such restriction.

The moral of this introduction (originally printed on a parallel Earth): Science fiction is driven by fear or hope, while science is driven by necessity or curiosity. The overlap between their motivations is huge.

I am going to conclude this introduction by listing Arthur C. Clarke's three laws of prediction.[4] Popular works of science fiction from *Doctor Who* to *Star Trek* have cited his third law ad nauseam. I encourage you to embrace all of them, but for fun, try living the second:

1. When a distinguished but elderly scientist states that something is possible, he is almost certainly right. When he states that something is impossible, he is very probably wrong.
2. The only way of discovering the limits of the possible is to venture a little way past them into the impossible.
3. Any sufficiently advanced technology is indistinguishable from magic.

Buckle up, because here we go.

CHAPTER 1
ONCE UPON A SPACETIME

Logic will get you from A to Z; imagination will get you everywhere.

—Albert Einstein

Science fiction books, movies, and television shows get a lot of mileage out of driving their characters through space and through time. In some cases, the hero manages to travel through both with the benefit of a Time and Relative Dimensions in Space (TARDIS) machine. This is the space and time vehicle of choice in the television series *Doctor Who*.

They might also journey via the time dilation that affects rapid acceleration ships as seen in Orson Scott Card's novel (and 2013 movie) *Ender's Game*. This type of travel is possible, but the truth behind traveling through time and space is much simpler. You don't actually need fancy technology because you are always traveling through time and space. If you jog a distance of five miles at an average speed of five miles per hour, then you have moved five miles through space and one hour through time.

I know, I know, this is obvious. But I want to point out that space and time are so intertwined in the physics of the universe that they must *always* be considered together. Physicists came up with a creative name for this unification: spacetime. The unification is also sometimes referred to as the space-time continuum.

Because of time's connection to space, scientists no longer believe, as Isaac Newton did, that time is absolute and flows uniformly. Albert Einstein's theory of special relativity proved that time is woven into the fabric of space and, oddly, is connected to the speed of light. This was a wow concept, but what did he mean by relativity?

If you are sitting at home reading this book you might feel stationary. This is an illusion. The earth spins as it circles the sun. The sun is traveling within the Milky Way galaxy, which itself is rotating around a supermassive black hole called Sagittarius A* (pronounced Sagittarius A-star). The Milky Way also moves within the local cluster of galaxies.

Ever since the big bang, space itself has continued expanding in all directions. So, every point (location) you can think of can be considered the center of our growing universe. There is no absolute position anywhere. No place within the universe is stationary; all movement is relative to something. Everything depends on point of view.

Think, for example, of a baseball game. The pitcher throws a fastball, and the batter takes a swing. From the bat's point of reference, the ball is moving toward it. But from the ball's reference, the bat is approaching.

An untrue tale: A physicist stood on one bank of a river, and Einstein stood on the other. "Hey," the physicist called. "How do I get to the other side?" Einstein thought for a few moments as he puffed on his pipe. Finally he called back, "You *are* on the other side!"

Einstein understood that almost everything is really a matter of perspective, the exception being the speed of light. That is absolute.

I wonder what science fiction author Theodore Sturgeon might have thought about this. According to the law named for him, *nothing is always absolutely so.* Then again, he wasn't a scientist (or, dare I say, an Einstein). Sturgeon is also credited with having said that ninety percent of everything is crap.[1] I promise you, special relativity is 100 percent crap-free. It has been proven experimentally and the results replicated. That's science.

With this in mind, the easiest way to reduce special relativity into something easy to learn (although granted not necessarily the most intuitive way) is to understand that everything moves at the speed of light. Yes, that's a weird idea, but our velocity through time added to our velocity through space *always* equals the speed of light.[2]

Some cool math hangs around behind this, but rather than blinding you with flashy equations, I'll tell you their conclusion: the faster we travel through space, the slower our journey becomes through time. In other words, as our speed increases, time bends to conserve the speed of light. This time bending is called *time dilation*. Don't worry, at our earthly speeds the time effects are minimal. Only after we have accelerated our starships to high speeds do the relativistic effects on time become more severe.

According to special relativity, as we travel through spacetime, we take our frame of reference—clocks—with us. On Earth, our personal clock is mostly in sync with everyone else's because of our slow speeds relative to each other. Now, if we board a starship and accelerate at great speeds away from our home planet, our clocks begin to differ from those we left behind. As our speed through space increases, we must travel more slowly into the future to conserve the speed of light.

In the vacuum of space, the speed of light is 299,792,458 meters/second (670,616,629 miles/hour).[3] If anyone ever asks, just say, "About 300,000,000 meters/second." It's easier to remember, and you'll look really, really smart.

Relativistic time travel is a great scientific tool for the science fiction toolbox. When the science in fiction is accurate, characters traveling at high velocities through space should be experiencing (suffering from, really) time dilation. Two examples of relativistic time travel in science fiction stand out. The first is *Forever War* by Joe Haldeman;[4] the second is Orson Scott Card's Ender series (known as the Ender Quintet).[5]

In *Forever War*, time dilation pops in when troops travel to military encounters on different planets. Each battle takes our hero William Manella centuries farther from the earth he knows. When eventually he returns home, he is so socially displaced that his language is archaic and his heterosexuality is repulsive. Imagine two races fighting each other hindered by time dilation, unaware of the enemy's stage of development when they next engage. It really is all relative!

The Ender series takes place over centuries, which are merely decades to Ender who travels at relativistic speeds. Much of the Ender universe revolves around a relativistic war with aliens that humans have charmingly named "buggers." A hero from an earlier confrontation with the buggers has been stashed away in the spaceship (eighty years before the book's setting). He is sent on a journey at near light speed that will return him to Earth when the humans have their fleet ready for the final battle against the buggers who have also been taking advantage of time dilation on their (relatively) long journey into human space.

EINSTEIN CONSIDERS THE GRAVITY OF HIS IDEAS

When we add gravity to our relativity soup, we get something that tastes a bit different than special relativity. This is Einstein's theory of general relativity.

Imagine you are in an elevator that is descending very quickly but not terminally. You should feel lighter, as if less gravity was pushing down on you (yes, that's right, pushing; we'll get to why later). Einstein used complex geometry to show how gravity and acceleration are equivalent.

Consider twins named Alice and Betty who both volunteer to be blindfolded and gently knocked out for science. Don't ask. I can't explain their motives.

Anyway, Alice wakes up in a spaceship accelerating from the planet at a rate of one unit of Earth gravity (1g), while Betty wakes up in an empty back room of a Starbucks in Flushing, New York. They both push themselves up and are asked if they know whether they are in a spaceship or on Earth. Neither will know for sure, except Betty thinks she smells coffee. They don't know because they feel an equivalent amount of force pushing down on them. Einstein concluded that the reason we feel gravity is because we are always accelerating.

Quick fact: the gravity of Earth (g) = 9.8 meters per second every second = 32 feet per second every second.

It might help if we bring general relativity down a couple of dimensions. Imagine a two-dimensional sheet pulled taut with a bowling ball sitting at its center. The weight of the ball curves the sheet around it. The area closer to the ball will have a steeper dip than areas farther away. Objects such as marbles will tend to roll toward the dip and then circle around the bowling ball.

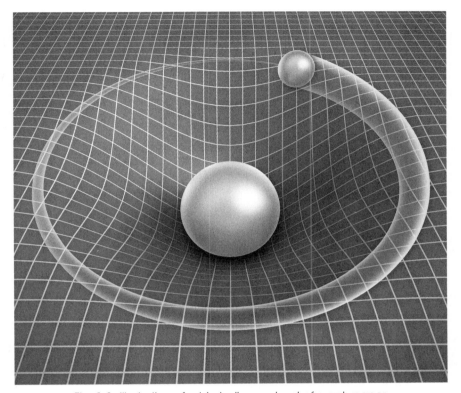

Fig. 1.1. Illustration of a big ball on a sheet of graph paper. (iStock Photo/koya79.)

Even light, like a marble on our sheet, will roll around massive stellar objects. General relativity teaches us that masses don't pull on each other. Instead, the force makes mass fall along curves. I believe physicist John Wheeler summed up the geometric nature of general relativity best when he said, "Spacetime tells matter how to move, but matter tells spacetime how to curve."[6]

Technical note: The sheet metaphor is incomplete because the fabric of spacetime is not a flat stretched sheet. It is shaped in such a way that the earth presses on it from all directions. This is some complicated geometry.

To me, the most striking consequence of rolling in these curves is *gravitational* time dilation. In our four-dimensional universe, massive objects such as planets or stars not only stretch and dent space, they also stretch time. Yes: the steeper the curve in spacetime, the slower time flows.

This is consistent with special relativity. As gravity increases, you fall faster (acceleration). When stuff accelerates, time slows (time dilation). The larger the mass, the more stretched and warped spacetime becomes, and the slower time flows for anyone near the surface. Therefore time does not pass at the same rate everywhere. It is flexible. Our modern satellites use Einstein's field equations to compensate for this. Otherwise the GPS for our cars and smartphones wouldn't be accurate.[7]

Did you know that the earth's core is about two and a half years younger than its surface? General relativity explains this too. Gravity at the earth's core is greater than on the surface, meaning that clocks will tick a bit slower down below. Over the 4.5 billion years of the earth's existence, the lag between the surface and core clocks have added up to about two and a half years.[8]

Something to ponder: do you need gravity to have time?

A movie example is *Interstellar*, which was based on exchanges between physicist Kip Thorne and producer Lynda Obst. This movie is a love letter to anyone who loves hard science. Tons of time dilation problems need to be overcome in this movie. Oh, and a lot of general relativity is included as a bonus. If you ask me, gravity is a supporting character. Joseph Cooper, the main character, spends a short time (relatively speaking) in the gravity well of a black hole. Eighty years have passed on Earth by the time he returns. Oops.

MASS IS ENERGY, AND SOMETIMES IT MAKES HOLES

In addition to the gravity-acceleration equivalency principle, Einstein's general relativity equations show that the geometry of spacetime is equal to the energy of all the stuff in it. The stuff in this context can be either mass or energy. In fact according to Einstein, gravity moving matter (and energy) along curves is what embodied spacetime.[9]

I'll let you in on something that isn't a secret: mass and energy are different forms of the same type of stuff. Einstein proved this when he derived his famous equation, E=mc² (energy equals mass multiplied by the speed of light squared). In non-math language, this means that the mass of an object is the measure of its energy content. A lot of energy is trapped in the mass of your body. (In a chapter bonus, I'll tell you how much.)

Because of the energy-mass equivalency, as an object accelerates, its mass increases. How? The answer is special relativity. As an object increases velocity relative to an observer, the amount of energy it carries also increases so it will appear more massive to the observer. In fact, the energy (mass) it would take to accelerate a single particle to the speed of light approaches infinity. You can thank the properties of spacetime for the speed of light limit.

If we push general relativity to its extreme, we arrive at a black hole, a location in spacetime that is infinitely warped by a singularity. A *singularity* is an infinitely small and dense point, something that is not clearly defined in physics.

In fact, this is where physics ends because infinity is not a number. It is an idea. As such, it can't be used in mathematical operations. And yet, black holes exist. Their gravity is so intense that, according to general relativity, time must slow to a stop. Black holes are covered in more depth in chapter 4.

NO WITNESSES . . . YET

Here is a question I imagine you desperately (I have a big imagination) want to ask: if, according to general relativity, time travel is theoretically allowed, then outside of science fiction, why haven't we met time travelers?

Think of time travel as a journey along a two-lane highway that contains many on- and off-ramps. As a time traveler, you can't visit an earlier era unless an exit ramp to that era exists. This gives us the following scientific rule of time travel: it is *impossible* to travel back to a time before time machines exist. This rule comes from the equations for general relativity where solutions for the continuum must exist at both ends, no matter whether we are talking about location or time.

But, for the sake of argument, if an alien race invented a time machine, say, a thousand years ago, and we got our greedy hands on the machine, it might be possible to travel a thousand years into our past. Sorry, *Back to the Future*, but your DeLorean time jump isn't sufficiently scientific for me. That didn't stop me from watching it three times.

TWO SCIENTIFICALLY PLAUSIBLE WAYS TO SURF SPACETIME WAVES

1. Wormholes (hyperspace in 4-D spacetime)

The idea of wormholes in space emerges from Einstein's theory that gravity arises from objects warping space. Wormholes are sometimes called *Einstein-Rosen bridges* because the idea of shared interiors was first fleshed out in a 1935 paper cowritten by Albert Einstein and Nathan Rosen. They showed that it was theoretically possible that the inward path of a black hole could be matched to a path that merged outward again at some different neighborhood of spacetime.[10]

It is not unimaginable to consider the possibility of two extreme local distortions warping space at different locations. If these bulges were to connect, you'd get a wormhole, a connection between the two distant locations. The series *Stargate SG-1* revolves around an Einstein-Rosen bridge portal device.

As any Trekie will happily remind you, a stable wormhole is the raison d'être of the television series *Star Trek: Deep Space Nine*. In the series, a stable wormhole connects the Alpha Quadrant, the location of the mainline characters, with the Gamma Quadrant—the home of a lot of unknown aliens.

The young adult (YA) story "The Outpost," written by Paige Daniels for *Brave New Girls: Tales of Girls and Gadgets*, takes place in a well-defined fictional universe where a group of profiteers known as Keepers control access to wormholes. This elite group controls access to interstellar travel, granting them control on all colony trade.[11] Who will stand up to them? There you go: plot plus science.

What if I want a wormhole that also integrates time travel? It is theoretically possible to use a wormhole as a time machine. Of course, it is a whole lot easier with the help of fiction than in real life because it involves moving one end of a wormhole.

Fig. 1.2. Illustration of a wormhole. (iStock Photo/eugendobric.)

First, we'd need a starship that can generate a gravity field strong enough to attract one end of a wormhole (remember: energy is connected to mass that is connected to gravity). Next, we'd have to drag the end along at speeds approaching the speed of light. As we know from special relativity, the clocks at the end we are dragging would run slower compared to the nonaccelerating end. Thus, we create a connection between two time zones where one end leads to the future and the other to the past.

Okay, let me combine a few ideas and add a dash of *Doctor Who* science to build a time and space machine. Of course, I'd have to come from a race that developed a time machine very early on because I can't violate the whole "not traveling farther back than the first time machine" rule.

To begin, I would need a black hole and the technology to

manipulate its gravitational waves. Gravitational waves as predicted by general relativity are ripples in spacetime that spread out from accelerated mass. We will go more in-depth on gravitational waves and black holes in future chapters, where they will take up both space and time.

Anyway, this setup would allow me to bend the on- and off-ramps of my time tunnel. Maybe, to make my life easier, I would lock a few fixed points into time for reference. Next, because it would take a lot of space to store my equipment, time rotors, power source, and so on, I'd need some sort of container to put all of this in.

Why don't I stow it in a higher-dimension tesseract? A *tesseract* is to a cube what the cube is to a square. It is folded space, dimensions within dimensions. This would allow me to store a lot of stuff inside something with a small volume (i.e., a container that is bigger on the inside than on the outside). Let's say my tesseract is the size of a box large enough to walk into. And why not paint it blue and call it a TARDIS?

In Madeleine L'Engle's *A Wrinkle in Time*, a tesseract is a time and space machine that is a "wrinkle" in space and time.[12] This is a fantasy device used to explain travel through space and time rather than explaining hard science, but it's a cool enough idea to mention. Come on, in what other fictional world can you travel by *tessing*?

2. Cosmic strings

This is a weaker theory than wormholes but still workable for science fiction. I once used it in a story to explain interstellar travel after I got a little bored with wormholes.

A cosmic string is a long, thin, and nearly infinitely dense theoretical remnant of the big bang. Consider it a long-line singularity. To build a space and time machine with them, you need to set two of these strings in parallel. Their relative motion, along with the geometry of spacetime, would allow for a closed time curve, allowing for backward time travel. However, as physicist Stephen Hawking has pointed out, most likely this system would collapse into a black hole.[13]

In the *Star Trek: The Next Generation* episode "The Loss," the *Enterprise* encounters a two-dimensional string.[14] Guess what? Mayhem ensues.

PARTING COMMENTS

We live in four-dimensional spacetime composed of three spatial dimensions (length, width, and height) and one dimension of time. The speed of light is the galactic speed limit of light, gravity, and time.

Time is not absolute, and it doesn't flow uniformly. Einstein showed us that time is woven into the fabric of space. His theories united space with time, matter with energy, and everything to gravity. From special relativity we learn that time is flexible and that mass has energy content. From general relativity we learn that gravity is not about pulling, but it is about falling, and if you add stuff to spacetime, you will change its geometry and fall differently (faster).

Fortunately for both scientists and producers of good science fiction, Einstein's theories left a lot of low-hanging fruit with space travel, time travel, wormholes, cosmic strings, and the existence of black holes all providing nibbles.

Something to ponder: In many scientific theories, time appears to emerge from the physical laws of our universe. Simply, our universe may bring time along for the ride as it expands. What about universes formed under different conditions? Do they have time?

CHAPTER 1 BONUS MATERIALS

Bonus 1: A Special Relativity Paradox

A paradox is defined as a statement or a collection of statements that might sound true but that lead to contradictions that defy intuition. Personally, I'm not convinced that paradoxes exist outside the rocky marriage of philosophy and mathematics that we call logic. Most of the paradoxes you'll come across are self-referential statements such as, "I always lie. Well? Did I or didn't I just tell the truth?" A paradox most likely means that something is wrong with the assumptions.

Nevertheless, we're going to fool around with things and ask one question: what do you get when you mix a sibling paradox with time travel? A wicked brew, I tell you. It is called the *twin paradox*.

I hope you remember the twins Alice and Betty because they're back. One day, Betty decides to board a rocket for a round-trip journey to Gliese 581, a star in the Libra constellation. Except for short periods of acceleration and deceleration near Earth and the star, Betty will travel at a constant velocity close to the speed of light. Gliese 581 is approximately twenty light-years away, so Betty's round-trip journey will take a little over forty years to complete.

Earth-homebody Alice knows all about special relativity (the faster through space, the slower through time), so she recognizes that time will run more slowly for Betty during her voyage. Alice calculates that Betty will actually have aged only four years when she returns.

It is all relative, so I will now retell the story from Betty's perspective. This is where a paradox (might) creep in. From Betty's point of view, she's the one remaining stationary, and it's Earth and Alice who move away at a rapid velocity and a later return. So, using the same logic as before, it's Betty who experiences forty years of time and Alice is the one who has experienced only four years.

When the twins are finally reunited, who has aged forty years and who has aged four years? If they are both right, you have a paradox.

Luckily for our sanity, there is a flaw in this thought experiment. The paradox relies on the assumption that the journeys taken by the twins are equivalent. They aren't. The difference comes down to acceleration. Whereas Alice remained relativity stationary on Earth, Betty undergoes three periods of rapid acceleration: the beginning, the middle (at the U-turn), and at the end of her journey. During each period of acceleration, she jumped to a different inertial time frame. Each jump took her farther from her sister's time frame.

The correct result is that Alice has aged forty years and Betty has aged only four years. No paradox.

Bonus 2: How Much Energy Is Contained Within the Matter of Your Body?

$E=mc^2$ shows that the mass (m) of an object is really the measure of its energy (E) content in terms of the speed of light (c). This equation is a result of special relativity, and it shows how much energy is contained in matter.

Consider a person who weighs 150 pounds. For this, I'm going to have to go metric on you. Physics is much easier using metrics. One hundred fifty pounds is about sixty-eight kilograms. Using Einstein's equation, this person has 6.12×10^{18} (the little number hanging near the 10 is called an exponent, and it is shorthand for how many zeros to add to the end of the number; in this case 10^{18} is shorthand for 10 followed by 18 zeros) joules of energy (68 kg x 300,000,000 meters/second). This amounts to 1.7×10^{15} watt-hours or 1.7×10^{12} kilowatt-hours or 1,700,000 billion kilowatt-hours.

That is a lot of energy. According to the US Energy Information Administration, in 2016 the United States produced, from all sources, a mere 4,079 billion kilowatt-hours.[15] So this 150-pound person contains more than 417 years of US energy production.

Bonus 3: Mutants

I will admit that this topic has nothing to do with Einstein's theories of relativity, but it does pop up occasionally in time-travel fiction. I'm talking about time travel as a genetic disorder or mutant trait. This is not science. There are no scientific theories of time being dependent on an individual's genetics (perception of time is another matter). That doesn't mean we can't enjoy it as fantasy fiction. Examples include *The Time Traveler's Wife* (chrono displacement) and Hiro from the television series *Heroes*.

CHAPTER 2
IF YOU ARE UNCERTAIN, CALL A QUANTUM MECHANIC FOR A FIX

> *The only thing that makes life possible is permanent, intolerable uncertainty; not knowing what comes next.*
>
> —Ursula K. Le Guin

> *He lies here, somewhere.*
>
> —Werner Heisenberg's epitaph

Welcome to the spooky world of quantum mechanics. Don't run away! I promise this chapter won't be too scary. But I warn you, although it will sound like fiction, it really *is* science.

Quantum mechanics is a lot like Alice's Wonderland, so I'm going to ask you to believe at least one impossible thing before breakfast. That impossible thing is the Heisenberg uncertainty principle, the lynchpin of quantum theory. The principle states that it is impossible to simultaneously know for certain both the current location of a subatomic particle and the path it has taken to get there.

More importantly, these two uncertainties cannot be reduced to zero together, meaning that the more you learn about one, the less you know about the other. This uncertainty is not from flaws in measurement but from quantum fuzziness due to the wave nature of particles (more on waves in a moment). This principle has been proven by experiment and describes the absolute limit to what can ever be known. No matter how much we struggle against it, there will *always* be uncertainty in the quantum Wonderland. In 1927, Werner Heisenberg published his work on this principle,[1] and in 1932—with certainty—he won the Nobel Prize in Physics.

Joke: Werner Heisenberg was pulled over for speeding. The officer asked him if he knew how fast he was going. Heisenberg replied, "No, but I know where I am."

The reason for all the fuzziness is that particles can also be waves. This is the so-called wave-particle duality. In fact, anything with mass, including you, can be described as a wave with a frequency. Frequency is the number of cycles of an oscillation. Think of frequency as the number of waves that pass a fixed place in a given amount of time.

In quantum mechanics, size matters. The larger the wave relative to the size of an object, the greater the uncertainty and fuzziness. The reason you don't have (much) uncertainty in your location or how you journeyed to your favorite coffee shop to read this book is because your velocity is too slow and your mass is too large to overwhelm your practical certainty.

However, as we tumble down the rabbit hole, shrinking until we're the size of the subatomic Mad Hatter, our wavelength relative to our size will become larger and larger. A wavelength is defined as the distance between successive peaks of a wave (think of water in a lake). Because our wavelength increases (relative to our size), we become fuzzy to anyone trying to find us.

This fuzzy probability cloud is similar to where an electron is as it orbits an atomic nucleus. In fact, Heisenberg (allegedly) claimed that it made no sense to talk about where an electron was or what it was doing between measurements.

For the record, it is impossible to shrink smaller than your constituent parts (atoms), but go with me on the whole Mad Hatter thing. I believe it will help you conceptually. Inside that fuzzy haze, where is the Mad Hatter? I know he has to be somewhere because I sent him down the hole. I will have to guess. I want it to be an educated guess, so I will search in the place where he is most likely to be.

The best way to do this is to use the Schrödinger wave equation.[2]

This equation describes reality in terms of probability. Although this is not a perfect analogy, imagine that I flip a subatomic-sized coin. While it is spinning, I don't know whether it will land heads up or tails up, but I do know the probability. There is a 50 percent chance of it landing heads up. If this were a true quantum-level event, we could say that the coin is simultaneously in a state of both heads and tails while it is spinning, a state called *superposition*.

In quantum physics, superposition is defined as the total of all possible states in which an object can exist. When the coin settles on the ground, the probability cloud evaporates and there is only one result: either a head or a tail faces up. Scientists call this the *collapsing of the wave function*. Probability fades into the deterministic world in which we live. This is true for all subatomic particles like electrons.

More generally, the wave function describes a probability wave where the peak is the highest probability of an outcome. Think of a bell curve with its rounded peak and, on either side of the peak, descending tails associated with lower probability. A particle can be found anywhere in the wave, and, at the quantum level, the particle is in all locations in the wave (superposition). Only when the wave function comes crashing down after an observation does all uncertainty fade. Presto! We have a classical particle with a known position and path.

So the measuring doesn't cause the uncertainty; rather, the uncertainty is because the position and momentum are *undefined* (fuzzy) until measured. The act of observing changes what is observed. Spooky. This interpretation is known as the Copenhagen interpretation of quantum mechanics. If you take this thinking to the extreme, then measuring things creates the reality we observe along with all past behavior.

This interpretation did not sit well with Einstein and Schrödinger. Schrödinger attempted to demonstrate its absurdity with his famous cat thought experiment.[3] Allow me to paraphrase it.

Consider a Cheshire cat in a box with a device connected to a hammer and a glass tube of cyanide. The device contains an atomic isotope that has a 50 percent chance of decaying within an hour. If it *does* decay, the hammer will drop, breaking the glass tube and killing the Cheshire cat. Because the isotope is atomic, it is subject to the spookiness (probabilistic nature) of quantum mechanics.

After an hour, the question is whether the Cheshire cat is dead or alive. What happens if you don't check? According to Schrödinger, the Copenhagen interpretation would conclude that as long as the box remains closed, the Cheshire cat is in superposition, both dead and alive. Only when the cat is observed as being either dead or alive does it actually become dead or alive.

WHAT IS A QUANTUM LEAP?

Imagine you have drawn a series of bouncing ball pictures in a flip tablet. As you flip through your book from back to front, the ball should appear to be continuously moving. But we know better. Each page is a discrete (meaning noncontinuous) moment of the ball's path. The movement from one page to the next is analogous to a quantum leap, a discrete movement from one spot to the next. The whole might appear continuous, but deep down at the *moment* level, it is not.

A quantum leap refers to an abrupt movement from one state to another. In the quantum universe, there is no in-between transition. For example, an electron is fuzzy (remember to think of it as a wave) in its orbit around an atomic nucleus. When it changes to a different energy state, it leaps from one orbit to another rather than move in a continuous fashion.

Whereas chapter 1's theories on relativity view the universe as continuous, quantum mechanics treats it as granular (the technical term is *quantized*). Here is another way to think about it. If you keep zooming in closer and closer on a picture, it will begin to look pixelated. The same is true for time. If you slow time down enough, movement might no longer appear to be continuous but rather would leap from one moment to another. No smooth transitions.

THAT SPECIAL MOMENT . . . WHEN YOU JUST KNOW

Decoherence is the term scientists use for the initial interaction of a quantum particle with its environment. This is the moment when its position, location, and other traits can be measured. In other words, you go from an undecided superposition to an ordinary position. It is the instant we open the box and observe the Cheshire cat, and she goes from superposition to her fate. Decoherence is not the cause of the collapse (that may be due to observation); rather it is the decay of coherence of the wave function as it leaks into our classical environment.

In the 2013 movie *Coherence*, a loss of decoherence leads to parallel versions of the characters leaking out into the story.[4] This is a cool use of quantum mechanics as a plot device.

Now that I think about it, we might not exist if no one is looking for us while we are in the quantum Wonderland. Worse, we might never have existed to fall down the rabbit hole in the first place.

In a story I wrote for *M-Brand SF* titled "Chronology," I ratcheted up the fuzziness of the uncertainty principle to the point where my human protagonist stayed in superposition, trapped in many different states (each state being a different character). I kept the reader guessing which character would be left when finally observed, and once observed, all the other characters vanished along with their histories. The big lie came from applying quantum mechanical weirdness to the macroworld, which can't be done outside of fiction. To maintain a semblance of plausibility, I kept all the other science in the story as accurate as possible.

In the Copenhagen interpretation, our measurements cause the decoherence, but other interpretations are available. One such alternative is the many-worlds interpretation, originally suggested by Hugh Everett in 1957.[5] It can be a fun way to invoke quantum physics in science fiction.

Remember Schrödinger's cat? Now imagine it's possible for both states (dead/alive) to continue after the box is opened. Each state causes the universe to branch.

Joke: Schrödinger's cat walks into a bar . . . and doesn't.

The many-worlds interpretation doesn't explicitly rely on decoherence, so there is no need for an observer. In this interpretation, every quantum event causes our universe to split into parallel universes. Each nonzero probable outcome is realized. When I flip my quantum coin, both outcomes occur. The universe has branched into two: one in which the coin comes up heads, and another where it lands tails up.

In the many-worlds interpretation, there is no wave function collapse. The wave is always standing, and the particle can be found in all locations, each in a different universe. Every outcome for every subatomic particle has occurred, meaning every possible atomic reaction has occurred, meaning every possible formation of molecules has occurred, meaning every possible *you* has occurred.

> *If quantum mechanics hasn't profoundly shocked you, you haven't understood it yet.*
>
> —Niels Bohr

Now that you know all about fuzziness, uncertainty, and the wave-particle duality—all of which, technically, are three sides of the same quantum coin—there is another feature of quantum mechanics that is popular in science fiction: quantum entanglement.

Quantum entanglement is when two particles are said to share the same quantum state despite their locations. Notice, I did not write *similar* states—I wrote the *same* state, and I meant it. Any change in one particle instantaneously translates into a change for its entwined partner no matter how much distance separates the particles. They can be apart the distance of a universe or the width of a strand of hair. This entanglement is what Einstein described as a "spooky action at a distance."[6]

REAL SCIENCE AND SCIENCE FICTION

Two practical applications for entanglement can and should be exploited more in science fiction: quantum communication and quantum computing.

1. Quantum communication

    Theoretically (which is good enough for science fiction), quantum communication allows for instantaneous communication anywhere in the universe. The important idea here is that instantaneous implies faster-than-light (FTL) communication.

    Wait. In chapter 1, didn't I write that the speed of light was the absolute speed limit? I did, and that is why this quantum application is so extraordinary.

    Measurement is the key to a functional quantum communication system. The act of measuring one entangled particle will force the wave functions of both particles to collapse. The spin of the measured particle instantaneously makes the other particle spin the same way. Now you have a quantum communication system translating spin into language.

2. Quantum computing

    Quantum computing also exploits the spookiness of quantum entanglement. The computers we are currently enslaved to rely on byte-sized strings of binary data. A byte is the smallest unit of working memory used to encode a single character of text or number. Bytes are traditionally made up of eight binary digits called *bits*. A bit can have a value of either one or zero.

    All of which is old school because a quantum computer uses quantum bits, commonly called (by computer geeks) *qubits*. The qubit is described in terms of probability, until it is observed. The values of one and zero are in superposition, and the probabilities of being either one or the other rise and fall with time. Sometimes the probability of a one is higher than zero, and sometimes it is less.

    So now you have a qubit representing a value of one or/ and zero, and everything in between. The entanglements between qubits allow their probabilities to mix. This is quite a big deal because quantum computers can perform operations while the qubits are in this state of superposition and entanglement.

    Why is all this helpful? After an inputted problem is broken down into small parts, massive amounts of calculations can run in parallel (all the parts are run simultaneously) where the sum of qubit probabilities can be used to determine the most likely answer to the problem. Quantum computers can solve very complicated problems relatively quickly, ones that today's computers would need hundreds of years to solve.

Finally, let's get crazy for a moment and mix quantum computing with the many-worlds interpretation. What about quantum computing between parallel universes, or using the computing power of other universes to solve complex problems in our own?

MIND-BLOWING QUANTUM TRIVIA

1. In quantum mechanics, particles don't have definite positions or velocities unless they are observed. To get weirder, in some cases, particles only exist as part of an ensemble of many particles. None have any existence on their own.
2. According to quantum mechanics, the past and future are indefinite and might exist as a continuum of possibilities.

PARTING COMMENTS

The Heisenberg uncertainty principle does not mean that accurate quantum measurements cannot be taken. It means that they come at a cost—the increase in uncertainty of something else. Wave equations in quantum physics capture both the wave(ness) and particle(ness) of a subatomic particle. The size of the wavelength dictates the range of possibilities for a particle's position. This has been proven by experiment, but there are different interpretations of what it all means. Two major interpretations:

The Copenhagen interpretation states that a subatomic particle can be anywhere along the wave front but collapses to a single point when observed.

The many-worlds interpretation considers a never-collapsing (standing) wave that holds many branches, wherein each spike (an area of high probability) is perceived with 100 percent probability by some observer.

Examples of practical applications of quantum mechanics in science fiction:

Quantum communication relies on particle entanglement for instantaneous communication independent of distance.

Quantum computers take advantage of both entanglement and superposition to increase their computing power.

CHAPTER 2 BONUS MATERIALS

Bonus 1: A Time-Travel Paradox

This is the so-called grandfather paradox thought up by French journalist René Barjavel in his 1943 novel *Le Voyageur Imprudent* to show how traveling back in time is impossible.[7]

Imagine that you kill your grandfather at a time before your father was conceived. Does this mean you wouldn't have been born? And if you weren't born, how could you have gone back in time to off your grandfather in the first place? According to Barjavel, you exist right now because everything in the past happened the way it did. By going back in time to change events, you are preventing your own existence. Ouch.

In science-consistent science fiction, an author can avoid the paradox by accepting the many-worlds interpretation. Each time a character travels to the past, he generates a new timeline. The original timeline of the character's birth, the one where his grandfather has not faced his grand-patricidal doom, still exists. But now there is a new timeline in which the character was never born. Paradox avoided.

In a story I cowrote with Dr. Susanne Shay titled "Mirror, Mirror," published in *Tales from Elsewhere*, we used a version of this branching theory. The story included gender swapping and a very confused protagonist.

There is a bit of good news, or possibly bad, depending on your point of view. The reasoning does mean there has to be branching if you travel into the future. This means there is no problem with a grandfather zipping into the future to off his grandson.

Bonus 2: Photosynthesis and Quantum Mechanics

Photosynthesis is the process by which plants take in sunlight and carbon dioxide and churn out oxygen and glucose. This process has a remarkable 95 percent efficiency rate. Researchers have calculated that the efficiency rate should be closer to 50 percent.[8] How did it get so high?

It is possible that the conversion energy is in a quantum superposition state and travels along all molecular pathways simultaneously. Once the quickest, most efficient path is found, the probabilities collapse and take that path. The best path is always taken, so efficiency is boosted.[9]

If this is truly due to quantum superposition then the question for scientists is, how are biomolecules able to exploit this quantum effect, and how can we copy it and make super energy efficient solar cells (our current versions operate at about 20 percent efficiency)?

Bonus 3: What Are Quantum Suicide and Quantum Immortality?

This comes to us by way of the many-worlds interpretation. It's similar to the Schrödinger cat thought experiment, only now you're the cat. This isn't one of my favorite tales, but it is informative in a very dark way.

A man sits with a loaded gun aimed at his head. Every minute, he pulls the trigger. The gun is connected to an atomic meter and will only discharge a bullet if an isotope has decayed. He is allowing the quantum world to decide his fate. Every minute, there is a 50 percent chance of decay. After the first minute, he pulls the trigger and the gun fires, and doesn't, and the universe splits to oblige both possibilities.

The branch in which he dies is done (terminally) and won't split anymore based on the gun (quantum suicide). In the universe in which he survived, he is unaware of his other version's death. For this survivor, there are still two outcomes for the next minute. There will always be a universe where he lives. It doesn't matter how many times he pulls the trigger, in that universe, the gun never fires.

This is called *quantum immortality*. It doesn't mean that he lives forever, just that he won't die from the gun firing.

Bonus 4: Time-Traveling Text Messages

In the previous chapter, I described the entanglement of particles that exist in the same time period but in different locations. This is called *spatial entanglement*. Quantum theory also extends to temporal entanglement, meaning the same location but different time periods. You don't have to assume events separated by time are independent. I encourage you to take a philosophical moment to consider this because at the quantum level, the future can affect the past.

A detector in the past could store the information of a particle and generate data on how it could be detected again in the future. At some time in the future, the first detector would be replaced by another one set in the exact same location as the first. You would need to account for planetary movement and so on, but let's not overcomplicate something that is already overly complicated. The second detector would receive the information sent by the first and effectively become entangled with the first.

Bonus 5: Subatomic Uncertainty in the Classical World

In classical physics, nothing is fuzzy. Whatever is measured always has definite, well defined properties, so any possible uncertainty must come from perturbation (a change) that occurs from the act of measuring.

Let's return to the subatomic Wonderland where the Mad Hatter is still the size of an electron. To observe the little guy, we need light. When a photon (a bundle of light; light quantized) hits him, it perturbs him enough to make him run off. Photons have large wavelengths, providing a lot of running room, so the chances of finding him are smaller than the chances of finding Alice after she so unwisely drank from an unidentified flask.

If instead of light we use high-frequency gamma rays, which have a small wavelength, they will hit the Mad Hatter like missiles. The point of impact gives us a good idea of where he was. Unfortunately, we've knocked him away in an unpredictable fashion.

That was then. Scientists now know that uncertainty really exists because of the fuzziness caused by the wave-particle duality of subatomic particles.

FIRST INTERLUDE
A TOUCH OF ATOMIC THEORY

You are mostly empty space. Yes, there is a lot of space between all those electrons in your body and the nuclei they surround. So much so that if all that space were eliminated, you would collapse to a size smaller than a freckle on an ant.

That isn't the only weird thing about atoms. When you touch anything made of atoms, you are not really touching anything. And yet, you can touch and be touched; hands don't pass through you. What you feel is the electromagnetic force.

Take this book, for example. The electrons orbiting the atoms of your fingertips feel only the repulsion from the book's electrons. What you feel is the repulsion force, but what you *think* you felt is at the discretion of your powerful brain. This is a good thing. Your hand probably does not want to build a molecule with the book, so it repulses it away.

Actually there is a bit of chemical bonding in the form of friction. You wouldn't want the book to slip from your hand. That said, atoms chemically joining is a good thing. It holds mass together.

If you cut a slice of bread with a knife, the knife does not actually touch the bread. The atoms of the knife push aside the atoms of the bread.

All of this is thanks to electromagnetic force, one of the four fundamental forces of the universe. These forces are responsible for all the interactions of the Standard Model of particle physics, the model that to date is the best description of how the building blocks of matter (particles) play along with each other in our universal playground.

THE FOUR UNIVERSAL FORCES

- **Electromagnetism**. This force holds substances together, including atoms. Oh, yeah, and thanks to electromagnetism, let there be light. Everything we've seen of the universe comes from electromagnetic waves.
- **Gravity**. Some people find gravitational interactions attractive. Although gravity is in the Standard Model, it cannot be explained by it.
- **Weak nuclear force**. This force is responsible for radioactive decay. It sounds boring, but without it, no sun; therefore, no you.
- **Strong nuclear force**. This force holds together the nuclei of atoms. If not for this force, the positively charged protons within the atoms of your carbon-based body would repel each other. Thankfully, the strong nuclear force binds protons and neutrons together in the nuclei of their atoms.

Technical note: Each of the four forces has a carrier particle. These are particles that act like messengers between other particles. For example, a photon (a quanta of light, which is the term for a particle of light) is the messenger for electromagnetism. When two electrons get close, they send each other photonic "keep away" messages. The messages are forceful and push apart the electrons.

If a force particle exists for gravity, it travels incognito. Or worse, it's in witness protection and hidden away from scientists. And yet, to fit into the Standard Model, it must exert its force on all matter. There are whispers that it might be a graviton, the quantized embodiment of gravity. The graviton is the holy grail of quantum mechanics. If it ever comes out of hiding, scientists might finally be able to reconcile relativity with quantum mechanics.

A TOUCH OF ATOMIC THEORY   55

WHAT IS AN ATOM?

The answer might not be as simple as you might imagine. Back in the day (around 465 BCE), Democritus talked about how everything we observe is composed of indivisible particles called atoms (a word derived from the Greek *atomos*, which means "indivisible"). He believed that if you kept cutting something in half, there must come a point where you will not be able to cut it anymore.

These indivisible elementary particles are what make up everything around us. Given what we call an atom today, Democritus was both right and wrong. He was correct that everything is made up of atoms. But alas, they are not elementary because they can be broken down further. As you might have learned in middle school or high school (or earlier), an atom can be broken down into negatively charged electrons, positively charged protons, and neutrally charged neutrons.

All normal atoms have a neutral charge, meaning they all have the same number of electrons as protons. Atoms that do not have the same number of electrons relative to protons are called ions. The different charges of various ions cause atoms to combine into molecules. That's chemistry.

Now, things can get weird even here. Consider a helium atom. This little bit of "something" comes with two negatively charged electrons orbiting two positively charged protons. (Because of this, the atom has a neutral charge. The two negative electrons cancel the charges of the two positive protons. Their charges arise from the electromagnetic force.)

Have you ever heard about opposites attracting? That's how the electromagnetic force likes to roll. A positive charge can't resist the seductive attractiveness of a negative charge. Also, the like minds of two negative charges can't stand to hang out together. According to this intuition, a helium atom doesn't make sense. First off, from what we know of magnets, shouldn't the electrons crash down upon the positively charged protons? And second, why don't the two positively charged protons repulse each other?

The answer to the first question is that an electron's orbit doesn't decay because of wave-particle duality. If you don't remember this topic, take a refresher peek at chapter 2 where we learned how everything contains its own wave with a frequency. For a mental picture, think of the wave as a spring (from the side, it would look like a wave). As an electron gets closer to its nucleus, the spring grows tighter until there is a width at which it cannot be compressed any further. This keeps them from plunging into the nucleus.

The answer to the question of why protons in a nucleus stick together is the strong nuclear force. At very short distances, say the width of an atomic nucleus, the strong force is stronger than the electromagnetic force trying to separate them. In the interlude bonus, you learn how strong the forces are relative to each other.

HOW OLD IS THE OLDEST ATOM?

The answer is about 380,000 years after the big bang. The first atoms in our universe are hydrogen and helium atoms, the lightest of the atoms. The heavier ones did not come around until about 1.6 million years later when the first stars formed. The newbies, the heavier atoms (elements) such as iron, did not come around until after the first supernova. The larger the atom, the younger it probably is. This is because they are formed from the fusion of lighter ones.[1]

WHAT'S THE MATTER WITH ANTIMATTER?

Antimatter is matter, only different. It's different because it's made of bizarro particles that have the opposite electric charge from ordinary matter. In other words, the atoms inside antimatter have positive electrons, called *positrons*, hovering in an excited cloud above a nucleus composed of negatively charged protons called *antiprotons*. When a particle contacts its anti-twin, they annihilate each other. That energy has to go somewhere. In science fiction, this energy could be used as energy for engines or weapons.

Theoretically, for every bit of matter in our universe, there should be an equal amount of antimatter. And they should have eliminated each other . . . and yet, here we are. For some reason, in our neighborhood of the universe, matter edged out antimatter during the primordial era. This is not necessarily true in other regions of the universe. Somewhere out there might be an antimatter galaxy, which is something science fiction writers should consider. How about a starship coming across an antimatter galaxy, able only to watch and not touch?

The Nobel Prize–winning physicist Paul Dirac derived the equation that discovered antimatter in 1928.[2] This wasn't his intent, just happenstance, which is the motif of many discoveries. It all began when he had this really weird idea that the laws of nature should apply to all. Go figure. At the time, quantum mechanics as formulated by Schrödinger violated relativity, and relativity ignored quantum mechanics entirely. Dirac's equation successfully reconciled the two when he showed what happened to an electron traveling close to the speed of light.

The funny thing was that the equation also had a second consistent solution, the positron. In 1932, Carl D. Anderson discovered the positron, the first direct evidence of antimatter.[3] He won the Nobel Prize in 1936 for the discovery. Today positrons are already used in positron-emission tomography (PET) scanners.

How about something a little more down to earth? Although still in the concept stage for cancer treatment, using antimatter instead of traditional radiotherapy (which shoots X-rays or protons at a tumor) might be much safer for the patient. Antiprotons could be used to annihilate the protons in the nuclei of tumor atoms. Plus the release of energy would do even more damage to the tumor cells.[4]

GOING DEEP

Protons and neutrons are made up of quarks. Quarks come in what physicists call different flavors: up quark, down quark, charm quark, strange quark, top quark, and bottom quark.

To keep this interlude (almost) simple, I will limit the discussion to the up quark and the down quark because they are the most stable. Brace yourselves because I'm about to get a bit mathy, but it's nothing more than addition.

To review, a proton has a positive +1 charge, the electron a negative -1 charge, and the neutron a 0 charge. Now for the bizarre: the up quark has +2/3 charge and the down quarks are -1/3. Yep, these are fractional. I know, crazy. This isn't anything for non-physicists to worry about because, thanks to the strong nuclear force, they are never found alone in nature.

How quarks charge up the particles in an atom's nucleus:

1. A proton is made up of two up quarks and one down quark. Now for the math:
   +2/3 (up) +2/3 (up) -1/3 (down) = 1, a positive charge.
2. A neutron is composed of one up quark and two down quarks.
   +2/3 (up) -1/3 (down) -1/3 (down) = 0, a neutral charge.

To complicate this, no two-quark combination can get you to -1, 0, or 1. Well, almost no combination. For a two-quark combination to work, you need what is known as antiquarks (with -2/3 and +1/3 charges).

These next two paragraphs are for completeness so that you can nod the next time someone tries to impress you with nuclear physics.

Anything held together by the strong nuclear force is called a *hadron*. Any hadron that is made up of three quarks is called a *baryon* (such as protons and neutrons). Mesons, on the other hand, are hadrons made up of two quarks.

Electrons are elementary particles, meaning they can't be divided any further. They come from a different family unrelated to quarks, called *leptons*. Unlike hadrons, leptons do not interact with the strong nuclear force. Their moderator of choice is the electromagnetic force. Electrons are the lightest of the charged leptons. This is very cool for our existence because only the lightest of charged leptons are stable. It is this stability that allows chemistry to happen.

A TOUCH OF ATOMIC THEORY 59

FIRST INTERLUDE BONUS MATERIALS

Bonus 1: The Range of Influence of the Four Universal Forces

In both strength and range of influence, these four forces differ greatly. The strongest of the forces is (surprise!) the strong nuclear force. For reference, imagine the strength of the strong nuclear force defined as equaling one. The next strongest force is electromagnetism, which has only a strength 1/137 of the strong nuclear force. Then comes the weak nuclear force at an astounding 0.0000001 of the strong nuclear force. Finally, the weakest of all, gravity, is 0.00000 000000000000000000000000001 of the strong force.[5]

Yes, gravity is relatively puny. Our entire planet is pulling down on you, but all it takes is a kitchen magnet to make a paperclip jump off the ground. And you must have noticed how a wee bit of static electricity makes a piece of paper stick to your hand and overwhelms the entire planet's gravity exertion.

Both gravity and electromagnetism have an infinite range of influence, the strong and weak nuclear forces, less so. The strong nuclear force's influence is not more than the width of an atom's nucleus (0.000000000000001 of a meter, or 1 femtometer). The weak force is limited to about one-tenth of 1 percent of the diameter of a proton.[6]

Bonus 2: Nuclear Fusion versus Nuclear Fission

Fission is when heavy atomic nuclei release energy upon splitting. Fusion is from combining nuclei and moving up the periodic table of elements into heavier atoms. For example, in the early universe, hydrogen atoms fused together to form helium atoms. When two atoms fuse, the mass of the new atom is less than the sum of the two original atoms. The missing mass becomes energy via $E=mc^2$.

As a power source, the advantage of fusion is that no long-lived radioactive waste is produced as what happens with fission.

CHAPTER 3
STRUMMING OUR WAY INTO EXISTENCE

The universe as a giant harpstring, oscillating in and out of existence! What note does it play, by the way? Passages from the Numerical Harmonies, *I supposed?*

—Ursula K. Le Guin, *The Dispossessed*

Chapters 1 and 2 described the twin pillars of twentieth-century physics: quantum mechanics and Einstein's theories of relativity. Both pillars have been proven valid through experimentation and observation. Any discrepancies come from extreme cases, such as when subatomic particles encounter the crushing gravity of a black hole.

These discrepancies, mostly due to that rascal gravity, have pressed scientists to search for a grand unified theory that will reconcile relativity with quantum mechanics. Superstring theory (aka string theory) is one candidate for the theory of everything (ToE). It proposes that tiny strings vibrate everything into existence. These strings are so small that atoms seem really, really huge in comparison.

As the Doctor (from the television series *Doctor Who*) might say, string theory is intuitively "wibbly-wobbly timey wimey"[1] and possibly more allegory than science. And yet, it might theoretically explain phenomena that can't be explained using the more conventional models described in the first two chapters. When stretched to its limits, string theory is compatible with many more forms of nature than are observed or predicted by conventional models.

There is no evidence for the existence of strings; however, it is based on solid (albeit complicated) math. This is a touchy topic among scientists because, yes, this math is able to describe the structure of nature, but it is also compatible with describing the natural world. This means that, according to math, all of these worlds must exist even though we cannot see them.

Now, if these universes somehow do exist, there is no causal contact between them and our universe. This is not science. So as long as strings are not observable and no direct experiments can test the theory, it must dwell more in the realm of philosophy than science.

EXTRA DIMENSIONS

String theory relies on the existence of dimensions that we cannot see or conceptually imagine, and, as I suggested before, math does the heavy lifting. Don't worry, you won't see any equations in this chapter!

As mentioned in the first interlude, the Standard Model explains most of what we observe in the universe but not everything (i.e., gravity). So the Standard Model is not quite a theory of everything; rather, it is a theory of most things. Some physicists believe we can go deeper than quarks (the constituent parts of protons) and electrons. To do this, we must depart the known and head for the speculative.

This is where string theory comes in. Instead of thinking of quarks and electrons as single-dimension particles, string theory suggests they might really have two or more dimensions. These dimensions might be tiny, curled-up ones, or so large that our three-dimensional universe can comfortably dwell within it.

Imagine the four strings of a violin. Each string is tuned (stretched) differently so that when they are bowed (an excitation), a different musical note is produced. This isn't too much different in string theory where the elementary particles (quarks, electrons, and their antimatter equivalent siblings) are the musical notes of strings. However, unlike our violin, which anchors the strings so they can stretch in different ways, the strings in the theory float in spacetime. They are tied to nothing, and yet they have tension.

Something to ponder: if they exist, where do the strings come from?

STRUMMING OUR WAY INTO EXISTENCE   63

A violin's music comes from its strings vibrating in three dimensions. When we draw these vibrations on a two-dimensional sheet of paper, it looks like a sine or cosine wave (math terms for wavy lines drawn across a flat screen or paper). The strings in string theory are strumming their music in ten, eleven, or twenty-six dimensions.[2] The fundamental particles in the Standard Model arise from these vibrations.

Chapter 1 described four of these dimensions: three spatial (length, width, and height) and one of time. The other six or seven or more, if they exist, must be hidden. Otherwise we would be able to experimentally detect them. A good hiding place would be to compact them to a size that is so small that they become Planck-length small—a millionth of a billionth of a billionth of a billionth of a centimeter. Named for Max Planck who defined the base units (length for example) used to define quantum mechanics, Planck-length is so small that classical ideas about physics are no longer valid. Quantum mechanics dominates.

I know this is hard to imagine, so let me help. Consider the edge of a piece of paper that is one millimeter thick. Now imagine a character named Ralph. He is insecure because of his size. He stands one-tenth the height of the paper's edge (0.1 mm). If his size were to represent the size of the entire observable universe, then Planck size would equal 0.1 mm relative to him.

In the early 1990s, physicists realized that string theory faced an uncomfortable dilemma: there was no single string theory. Five unique versions each successfully describe phenomena under certain conditions, and each theory requires an additional dimension or two to describe a particle in the Standard Model, but each breaks down while explaining other particles. If only the five could be united into a single theory then almost everything could be described.

This is where M-theory comes in. M-theory gives us an explanation for why so many dimensions are necessary. It treats each of the five string theories as subsets and serves as a road map to connect them. Of course, another dimension had to be added for M-theory to work. But who's counting? Don't ask me what the M stands for. It is a mystery in physics. I have heard many suggestions but nothing conclusive.

If there are fewer dimensions than quantum events then negative probabilities must be included. Trust me, if this is true then things get ugly. Scientists do not like ugly. So it is better to add them in than subtract them out. If extra dimensions do exist, they might be really small and rolled up into their space. Or possibly the extra dimensions might be very large and contain all of matter and gravity within them.

These large ones are called *membrane dimensions*, sometimes called *branes* by physicists. In brane theory, our three-dimensional universe might be a stretched brane floating through a four-dimensional background called the *bulk*. Imagine a two-dimensional sheet of paper riding the winds of our three-dimensional world. Add a dimension to both (along with a few other considerations), and you get a brane floating in the bulk.

Within the M-theory framework, a brane is required as an attachment point for all of the strings. The tiny dimensions are squashed down into a particular dimensional shape called *Calabi-Yau space*, from which they are able (mathematically) to produce all of the physics we are able to see. An atom's fundamental qualities depend on this geometry.

Science fiction has plenty of room for these extra dimensions. In Liu Cixin's novel *The Three-Body Problem*, Earth is invaded by technology hidden in curled dimensions.[3] China Meiville's *The City & the City* deals with a conflict between overlapping dimensions.[4] *Sunborn* by Jeffrey A. Carver uses a lot of the science ideas in this book. He has ancient AIs living in compact dimensions inside a black hole.[5] Now that you know all about hidden dimensions, all you need to do is read the chapters about black holes and AIs.

MEET THE COMPETITION

Loop quantum gravity (LQG) theory is the chief competitor in the search for a grand unified theory. Where string theory attempts to explain everything in the Standard Model and bring gravity into the family of universal forces, LQG is much more modest. It seeks only to reconcile quantum gravity with spacetime.

General relativity treats gravity as a property of the geometry of spacetime, while quantum mechanics treats gravity as a quantum force. LQG theory holds that spacetime itself might be quantized,

STRUMMING OUR WAY INTO EXISTENCE 65

meaning it treats space as granular rather than continuous as Einstein believed.

So, if you kept zooming in on an area of space, say the distance between you and this book, with an impossibly powerful microscope, you would begin to see space itself pixilate and appear granular. The theory holds that these grains are woven together by finite loops of gravity. This is profound because it means space might be discrete (individual grains) and not continuous.

Unlike string theory, there might be a way to test for loop quantum gravity. All you need to do is study the radiation a black hole evaporates. Researchers believe that if quantum gravity exists, measurable discrepancies will appear in the types of radiation evaporating from a black hole.[6]

One of the biggest challenges for researchers is to find an evaporating black hole. So far, no one has detected one. The same technique might also be used to find evidence of quantum gravity in background radiation left behind after the beginning of the universe. Don't worry if you have no idea what black hole evaporation means. The topic is absorbed into chapter 6. For now, just know that this theory is testable.

CAN QUANTIZED SPACE SOLVE A PARADOX AND HURT A VILLAIN?

Loop quantum gravity theory might answer the paradox of infinite distance. Allow me to unjustly turn you into a criminal mastermind having a bad day. You, the criminal, spot the Green Arrow, a hero of DC Comics, just in time to see the arrow launching toward your head. Greeny is in a take-no-prisoners state of mind.

If we believe Zeno of Elea (who lived during the 400s BCE), you are safe.[7] The arrow passes halfway across the warehouse you call a lair, then half the remaining distance, then half again, and so on. I've divided its journey into infinite numbers of shorter and shorter segments. A half, then a half of the half, then half of the half of the half, and so on. The arrow never hits you because it must pass through an infinite number of points that make you infinitely far away. The arrow always gets closer but never strikes. Math has saved you!

Only it won't. Not really. According to physics, you are about to feel a sharp pain and then probably nothing ever again. I will tell you why. There are a few philosophical and mathematical explanations I could provide, but let's go with the quantum one. At any given time after the shot is taken, only a *finite* number of quantized grains of space are between you and the arrow. Sorry, infinity is not going to help a criminal.

PARTING COMMENTS

The idea behind all the different versions of string theory is that strings are the most fundamental unit of nature. They strum their music in dimensions of the universe we cannot perceive, and yet they create all that we can see. To us, these dimensions might only exist as mathematical constructs. The tiniest of them might be curled into the tiniest of scales, which is Planck length. This size is so small that length might not matter anymore. The largest dimension might be a brane that contains our entire universe.

The five string theories are internally consistent, but separately they fail as an explanation of everything. M-theory is an umbrella theory that unites them. It is a road map for which theory is best at explaining which type of phenomenon.

Loop quantum gravity theory takes the more modest approach of not attempting to explain all particles. Instead it focuses only on gravity. If it can connect gravity to quantized spacetime, then it will have unified relativity and quantum mechanics.

FOR THE RECORD

My favorite intersection of Murphy's Law and string theory: anything in string theory that theoretically can go wrong will go wrong, but if nothing does go theoretically wrong, then experimentally, it is ruled out.

CHAPTER 4
OUR UNIVERSE (AS OPPOSED TO THOSE OTHERS)

If you wish to make an apple pie from scratch, you must first invent the universe.

—Carl Sagan

The science of studying the whole universe is called *cosmology*. Cosmologists are big-picture scientists. And, as you might have guessed, the universe is pretty big. It also comes with 13.8 billion years of history. This chapter tackles a few of the big questions in cosmology, including the universe's size and how it began.

HOW BIG IS THE UNIVERSE?

The answer: pretty big.

Okay, to start things off, start thinking big . . . really big, because the observable universe is about ninety-three billion light-years in diameter. A light-year is about six trillion miles. I will let you in on something extra: tomorrow the universe will be bigger.

How much bigger?

The universe is growing at approximately seventy kilometers per second per megaparsec.[1] This complicated-sounding rate is known as Hubble's constant. Allow me to attempt to untangle it for you.

Astronomers use parsecs to express stellar distances. A parsec represents 3.26 light-years. A megaparsec is a million parsecs (3,262,000 light-years). By the way, parsec is shorthand for *parallax of one arc-second*. If you are planning on writing your own space adventure, I suggest sticking to using parsec.

So if you peer out at the night sky using the powerful telescope NASA loaned you, at a distance of about 3.3 million light-years out into space, the galactic objects will appear to be receding from Earth at about seventy-one kilometers per second.

The deeper into space you set your telescope, the faster the expansion will appear. Look out far enough, and objects will vanish. The reason that distant areas of space appear to (and actually do) move faster than the speed of light (and vanish) relative to our galactic location is due to the geometric nature of the expansion.

The next time you watch *Star Wars: A New Hope* and Han Solo says, "You've never heard of the Millennium Falcon? It's the ship that made the Kessel run in less than twelve parsecs," remember that a parsec is a unit of *distance*, not one of *time*.[2] The only science fiction solution that might make this statement scientifically consistent is if Han meant that he was able to find the shortest route in and out of hyperspace.

Take a moment to appreciate how big the universe is and how fast it is expanding. Traveling at the speed of light, a photon (quantized light) leaving our sun takes a bit more than eight minutes to get here. To reach Pluto, the little guy needs five and a half hours. If the photon is interested in taking a road trip, the journey to the next closest star (Proxima Centuri) is roughly 4.2 light-years.

So, there is a lot of space out there in space. If we could travel at the speed of light (we can't), we would need over four years to get to our next stellar neighbor, Proxima Centuri. So, gallivanting around the Milky Way is time prohibitive. Given current technology, for the foreseeable future, humans will not make it far from Earth.

The news is not all bad. The science (if not the politics or the funding) exists now to create space stations or planetary bases in our solar system. If humanity is more ambitious, a lot of good ideas can be found for terraforming a few of the moons and planets in our solar system. This topic is covered in more detail in chapter 12.

If you imagine escaping the solar system, a sprinkle of science fiction dust might come to your rescue. In the Star Wars universe, ships travel in hyperspace lanes. These are wrinkles in spacetime that allow ships to jump from one point to another without traveling directly toward their destination.

In the Star Trek universe, warp drive technology, powered matter/antimatter annihilation mediated by dilithium crystals, is used. (See the first interlude for a definition of antimatter.) The best technobabble description of the warp engine is that it is a gravimetric field displacement engine powered by matter/antimatter reaction. This is supposed to mean that warp fields are generated around a starship to form a subspace bubble. The bubble distorts local spacetime, allowing the starship to slide through the distortion at velocities (warp factors) exceeding the speed of light. A more scientific description of a warp drive is found in chapter 17.

The speed of a warp factor is never clearly defined in any permutation of *Star Trek*. The different warp speeds do not make scientific sense to me, so I won't try to explain the exponential nature of the factors.

The best part: no time dilation when zipping around in either the Star Wars or Star Trek universes.

HOW IS THE OBSERVABLE UNIVERSE DIFFERENT FROM THE ACTUAL UNIVERSE?

When describing the universe's size, I deliberately used the term *observable*. I did this because the universe we live in is different from the universe we can see.

This is a big deal in cosmology. To astronomers, observable refers to the ability to see the light emanating from distant regions of the universe. However, some regions of the universe are so distant that the light from their stars has not had time to reach us yet. When it finally does, I suspect we will be long gone, but our descendants might get an eyeful.

Then there are other regions that are expanding away from us faster than the speed of light (according to general relativity, anything containing mass might not exceed the galactic speed limit while traveling through space, but this does not apply to space itself), so their light will never reach our pale blue dot. As the universe expands, the horizon will get smaller. The universe we can interact with will also get smaller.

The shrinking horizon isn't the only consequence of the galactic speed limit. It also means that nothing we see is current. The greater the distance, the greater the time differential. Time differentials are surprisingly commonplace and accepted in our perception of reality. I will not know what is going on in the world at the moment you read this sentence. Your now is my future.

When you look out at the stars, you see yesterday. Actually, what you see is yesterday's great-great-great-grandmother. The Milky Way is about 100,000 light-years wide, and our sun is pretty far out from the galactic center, so any light emitted from a star at the opposite end of the Milky Way would need 100,000 years to travel here. If you see it, then it is old light, a long-gone yesterday. Here's the kick in the pants: as far as the universe is concerned, this is more or less a recent snapshot. Any light you see from Andromeda, the next closest spiral galaxy, is about three million years old.[3]

Yes, it is sad that there are parts of the universe we'll never know about, but this isn't necessarily bad news for science fiction.

HOW ARE GALACTIC DISTANCES CALCULATED?

So, how do we know the distance between stars or between galaxies? Astronomers use cosmic yardsticks, also called *standard candles*, to measure distances. They also use a lot of geometry. For galaxies that are very far away, the yardstick is a supernova. Supernovas are ridiculously bright stellar explosions. Their observed brightness and a measurement of red-shifting due to expansion can be used to determine distance. More on red-shifting in a moment.

Closer to home, Cepheid stars are used to gauge distances. These are very luminous stars that pulsate (changing diameter, temperature, and brightness) in a predictable pattern. Henrietta Leavitt discovered the period-luminosity relationship in 1912.[4] Astronomers are able to measure distance because of the relationship between the brightness (seen by a telescope) and the pulsation period. Bonus 4 of this chapter provides a more detailed description of how Cepheid stars are used to measure distance.

AND IN THE (OR RATHER A) BEGINNING . . .

And now for an origin story . . . our origin story. I'm talking about the beginning of the observable universe, the big bang. The name is a misnomer because the event was neither big nor bangy.

The name started out as an insult by English astronomer Fred Hoyle made on BBC radio. He believed there was no such thing as a spectacular beginning. Instead, he believed in the competing theory called *steady state*. In that theory, the universe doesn't change over time, but the stuff within it, like galaxies, can move around. He was wrong, but he still gets credited for coming up with the name.[5]

So, what exactly is the big bang? It was an amazing event where all the matter that ever was and ever will be arose from a very (very) small point called a singularity. Recall from chapter 1 that a singularity is an infinitely small and dense point in spacetime. And 13.8 billion years ago, everything we observe spewed from one. Oh, and the big bang also created time.

You want evidence for a big bang? Good! You are thinking like a scientist. Below are three lines of evidence. There are more, but this is a good start.

1. Observation

    Through observation, Edwin Hubble discovered evidence for an expanding universe from the red-shifting of galaxies. When light from an object moving away from an observer shifts to the red end of the electromagnetic spectrum with respect to the observer, redshift has occurred. The color of the visible portion of the electromagnetic spectrum, in order of least to most energetic (frequency), is red, orange, yellow, green, blue, indigo, and violet. These are the colors of a rainbow.

    That shift of light to the red end of the spectrum is similar to the Doppler Effect. The sound of an object changes with its movement relative to an observer. As an ambulance rushes toward you, for example, the sound's frequency increases and pitch rises. After it has dashed past you, the frequency decreases and the pitch becomes lower.

    By studying the redshift of moving galaxies, Hubble proved

they were moving away from the Milky Way.[6] He also showed that the farther away the galaxy, the faster it appeared to be traveling and the younger it appeared. The really distant galaxies appear to have been formed recently or as glowing gas that has yet to develop into stars. This evidence, plus general relativity, allowed cosmologists to rewind cosmic history. They showed that the farther back in time you go, the smaller the universe was.

Bottom line, astronomers are able to see the observable universe spreading out. It should not be difficult to imagine that a year ago, it was a little smaller than it is today. If we wind the clock backward 13.8 billion years, the universe must have been only a primordial point in space.

2. Cosmic microwave background

The cosmic microwave background (CMB) is a baby picture of the universe. It is what the universe looked like when it was only 380,000 years old, and it represents the furthest back in time that we can peer.[7] The CMB is a thermal energy leftover of the big bang, a remnant, an afterglow.

What is so marvelous about the CMB is that its existence is predicted from the big bang theory and (wait for it) CMB is detectable. From every direction, when we look out at the galaxy, we see evidence of the universe's earliest light. Also, the CMB appears the same in all directions, meaning there is no up for the universe. It is isotropic.

Fig. 4.1. Illustration of cosmic microwaves. (NASA/JPL-Caltech.)

3. The elements (big bang nucleosynthesis)
    Our universe contains a lot of elements. For example, there is the carbon-based you reading a carbon-based book (or the composite material of an electronic device). These elements did not exist in the early period of the universe. When it all began, there was nothing more than hydrogen nuclei compressed into a tiny volume. About ten seconds to twenty minutes after the big event, the universe's hydrogen nuclei began fusing into helium.[8] Over hundreds of millions of years and a lot more nuclear fusion, the heavier elements were created.

Something to ponder: the universe began mostly as hydrogen that, after quite a lot of time (and by fusion and evolution), became people to think about what hydrogen actually is.

Based on the elements of the early universe, the big bang standard theory successfully predicts how much of the elements we should (and do) observe today. For example, our sun contains hydrogen and some helium, and yet we have heavier elements here on Earth. These came from previous generations of stars.

WHAT CAUSED THE BIG BANG?

Fair question, but I don't have a fair answer because the question itself might be meaningless. Consider how *cause* comes before *effect*. In the standard theory, there was no *before* (i.e., time) before the big bang. The question of time before the big bang is like asking what is north of the North Pole. As you read in chapter 2 (on quantum mechanics), some things happen randomly for no particular reason; they have no particular cause. This is a capricious property of the universe. Yes, there might have been some cause, but our scientific understanding of the universe doesn't require one.

A good book that blends fiction and nonfiction on this subject is *George and the Big Bang*.[9] It is the third book in a young adult science adventure series written by the daughter-father team of Lucy and

Stephen Hawking. In the story, there is a scheme to destroy the large hadron collider (LHC) before it can conduct an experiment to recreate the initial conditions of the big bang. Yes, mayhem and science ensue. The book also includes essays by Professor Hawking and other scientists about the origins of the universe.

A more speculative origin story proposed that the big bang occurred, but it wasn't the beginning of everything. I am talking about brane theory, one of the string theories described in chapter 3. It hypothesized that our three-dimensional universe is a stretched mem(brane) floating through a higher dimensional background called the bulk.[10]

Now imagine that our brane is not the only one floating through the bulk. Occasionally two branes collide, like two hands clapped together, releasing a lot of kinetic energy. To viewers inside the clapping branes, it would look like a big explosion, a big bang. The kinetic energy would create matter (you do remember the energy-matter equivalency from chapter 1, right?); matter would give rise to a universe, a universe might give rise to life, and life-forms could include creatures such as us science fiction geeks.

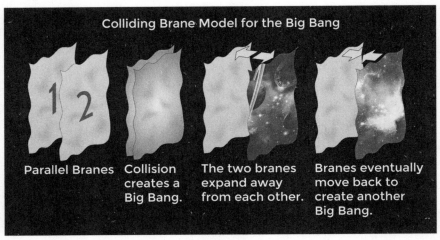

Fig. 4.2. Illustration of two branes colliding.

WHAT HAPPENED NEXT?

Let me guess: you've heard enough about the big bang. You want to know what happened next. I'm here to serve.

During the first few trillionths of a second after the big bang, the universe was so hot and compressed that spacetime boiled with energy. After that brief hotfoot, the universe made a phase transition from that high energy state to a lower one. A lot of energy fell out of the vacuum, accelerating the already expanding universe. This brief period of cosmic inflation was first proposed by Alan Guth in 1981.[11]

After 0.0001 of a second after the big bang, quarks got together to form protons and neutrons. It took between ten seconds and twenty minutes before the first atomic nuclei made an appearance and 300,000 years before electrons are captured into orbits of the atomic nuclei and the first complete atoms arise.[12]

Based on microwave background radiation emitted about 380,000 years after the big bang, cosmologists are able to piece together what happened next. And it is a tale of the Dark Ages. As the universe continued its expansion, it grew cold and dark. In the otherwise uniform distribution of the CMB is evidence of pockets of small-scale clumps.

This clumping would become the first small protogalaxies, masses of gas that form galaxies. These early protogalaxies are different from the ones astronomers see today. Back then, they were mostly only hydrogen and helium. Today they include heavier elements that are created from stars.

The first clumps capable of forming stars would arise between 100 million to 250 million years after the big bang.[13] The scaffolding on which these first galaxies would be built was dark matter. Let's face it: there wasn't enough ordinary matter in the universe to create the gravity needed for these gas clouds to form suns. In the Dark Age, dark matter and ordinary matter hung out together in the protogalaxies. Over time they must have had a feud because today, ordinary matter hangs out with us in the galaxy's inner region while dark matter hangs out in the galaxy's outer halo.

Let's go back to when they did get along.

The gas clouds began to compress and rotate under the weight of gravity from the dark and ordinary matter. The pressure caused heat to build up, and hydrogen atoms combined into hydrogen molecules. These molecules cooled the densest parts of the gas. As the rotating gas flattened and cooled, ordinary matter separated from the dark matter. The densest clumps of gas continued contracting until some of these clumps collapsed into stars. Yes! Let there be light. The end of the Dark Age.

Fear not, this isn't the end of the story for dark matter. The third interlude will try to light it up for you.

After the universal Dark Age, first-generation stars burned very hot because they lacked the heavier elements. It is more difficult to produce nuclear energy without the heavier elements, so a first-generation star had to be hotter to produce enough energy to counteract gravity.

These stars were a wild bunch. They burned bright and lived short lives that lasted only a few million years. By comparison, our sun is middle-aged at 4.5 billion years old. Some of the early stars eventually exploded and became supernovae that fused the heavier elements into existence. This is the birth of metal.

WHAT ARE GRAVITATIONAL WAVES?

This was already answered in chapter 1, but it is very much worth mentioning here because gravitational waves might tell us more about our history.

The math of general relativity shows that gravitational waves are disruptions in spacetime that cause ripples where time speeds up, slows down, and speeds up again. They might be caused by violent high-energy processes such as the big bang, or by massive objects like black holes or neutron stars moving through spacetime.

They carry information about their cause at the speed of light. These types of waves are believed to be unaltered as they ripple through space, meaning they don't fade away like a wave on a lake. This "unalterable" property might help scientists learn what happened before the earliest light (the cosmic microwave background).

As I've said before and will say again, science loves evidence. The

OUR UNIVERSE

theory of cosmic inflation suggests that there was a rapid expansion of spacetime after the first fraction of a second after the big bang. This sudden "push" might have produced detectable ripples that prove the theory. This wave may carry with it new information about this earliest moment.

HAVE GRAVITATIONAL WAVES EVER BEEN DETECTED?

Yes! Colliding black holes 1.3 billion light-years away gave researchers the perfect chance to confirm the existence of gravitational waves. The colliding black holes created a violent storm in the fabric of space and time. It swept across space until it reached the Laser Interferometer Gravitational-Wave Observatory (LIGO) detectors in 2015.[14] After confirming the results, the LIGO team formally announced the spacetime ripples in 2016—around the one hundredth anniversary of Einstein's theory of general relativity, which first predicted their existence.

The LIGO study used laser beams projected along pipes; the beams were reflected in mirrors at each end. Researchers monitored the distance between the mirrors for fluctuations as space expanded and contracted. The amplitude of the gravitational wave was smaller than the size of an atom. Nonetheless, the experiment further confirmed general relativity.

SOME GENERAL FACTS ABOUT OUR UNIVERSE (AS WE KNOW IT)

- 13.82 billion years ago (BYA), the big bang went bang.
- 300,000 years after the big bang, the first complete (with electrons and protons) atoms form.
- 380,000 years after the big bang the cosmic microwave background radiation is emitted.
- Two hundred million years after the big bang, the first stars ignited and there was light.

- The universe is ninety-three billion light-years wide and growing.
- A light-year is about six trillion miles.
- At the latest census, at least two trillion galaxies existed in the universe.[15]
- Our galaxy is called the Milky Way. It is 100,000 light-years wide and contains at least two hundred billion stars. It is about 13.6 billion years old.
- The Milky Way contains between seven hundred billion and one trillion solar masses. This includes all the stars, planets, moons, and just about anything that has mass. A solar mass is equal to the mass of our sun. The sun's mass is about two *nonillion* kilograms (that is two followed by thirty zeros).[16]

  To visualize a nonillion, imagine that you've counted every grain of sand on all the beaches and in all the deserts of Earth. You'd come to a nice round number of about seven quintillion, five hundred quadrillion. (You would have to set aside a very long day from your schedule for this task.) If you then added that number to the estimated number of stars in the Milky Way (two hundred billion), your count would still be less than half the value of a nonillion.[17]
- Up to about one hundred years ago, it was believed that the Milky Way was the entire universe. It is not.
- Our sun is about twenty-six thousand light-years from the Milky Way's center.
- The Milky Way moves through space at a velocity of about 550 kilometers per second. It rotates around its center at a rate of about 220 kilometers per second.[18]
- The Milky Way is a spiral galaxy dragging four major arms as it spins. Older stars tend to occupy the center hub while the newbies take up residency in the spiral arms. Earth is in the Orion Arm, which possibly is a bridge between the Sagittarius and Perseus arms.
- At the center of the Milky Way is a black hole with the mass of around 2.6 million suns.[19] This supermassive black hole is named Sagittarius A*.
- The Milky Way is part of a local cluster group of fifty-four gal-

OUR UNIVERSE   79

axies that are held together by mutual gravitational attraction. This Local Group is about one hundred million light-years across.
- Eventually the Milky Way will collide with its neighbor, the Andromeda galaxy. Don't worry. This won't be a problem for about four billion years.

OUR SOLAR SYSTEM, HOME OF SOL, OUR SUN

- Our solar system has eight planets and ten *confirmed* dwarf planets (the actual number of dwarf planets in our solar system is probably closer to four hundred). A dwarf planet is similar to a planet, but it is not large enough to gravitationally clear its orbit around the sun. Pluto is a dwarf planet because, with its partner Charon, it is part of a cohabiting binary system. They are tidally locked, meaning they always present the same face to each other as they circle each other. This union orbits the sun.

  To date, the farthest dwarf planet found on the outskirts of the solar system is called Sedna. The second-most distant dwarf has the cute name Deedee (for distant dwarf). The closest is Ceres, where evidence of water was recently discovered.[20]
- The sun makes up 99.9 percent of the mass of the solar system. It can hold around one million Earths.[21] Someday the sun will shrink to the size of the earth. After this slimming down (while retaining mass), our sun will be a white dwarf. This won't be for a while, because the sun is currently only middle-aged (at 4.5 billion years). Its core is about fifteen million degrees Celsius.
- It takes eight minutes and twenty seconds for the light of the sun to reach the earth. It takes 1.3 seconds for light from the moon to reach the earth.[22]
- An astronomical unit (AU) is defined as the average distance between the center of the sun and the center of the earth. This is about 150 million kilometers.
- Traveling outward within our solar system from the sun, we first pass the four smaller rocky inner planets (Mercury, Venus, Earth, and Mars). Next, we travel through an asteroid belt that

includes a dwarf planet rock named Ceres. After exiting the belt, we come to the four gas giants (Jupiter, Saturn, Uranus, and Neptune).
- Jupiter's gravity helps protect Earth from meteors. Think of it as a giant vacuum sucking up space debris. But I'm sure the dinosaurs will tell you that Jupiter doesn't get them all.
- The Kuiper (rhymes with diaper) Belt begins just past Neptune. It is disk-shaped and circles the sun at thirty AU out to about fifty-five AU.[23] It is an area of icy objects and short-period comets. Short period is defined as orbiting the sun in less than two hundred years. Pluto is part of the Kuiper Belt. For the record, Pluto is not a comet.
- Speaking of Pluto, did you know it has liquid water? About one hundred miles beneath its surface there is (probably) a slushy ocean that might be sixty miles deep. The dwarf planet has retained enough radioactive heat from its formation to keep the water in a semi-liquid state.[24]
- The Oort cloud is a shell of about two trillion icy objects in the outermost region of the solar system. Its innermost edge may be as close as two thousand AU from the sun and the outer edge as far as 200,000 AU.[25] Long-period comets might have their origins here. Long period is defined as having orbital periods longer than two hundred years.

PARTING COMMENTS

The universe has a history, and its structure is explained by the theories of relativity and quantum mechanics. The big bang theory is the leading theory of its origin, but it doesn't tell what came before or why it even had to begin. Quantum theory or string theory, on the other hand, might have something to say on this subject. The universe might come from quantum entanglement of quantized spacetime, or it might have been the result of branes clapping together in a higher dimension, or both, or neither. Cosmology is exciting that way.

CHAPTER 4 BONUS MATERIALS

Bonus 1: Olbers's Paradox

Named after Heinrich Wilhelm Olbers, this paradox asks this question: if there are billions and billions of stars, then why isn't the night sky completely lit up? If the universe is infinite and eternal, the night sky should be uniformly bright.

An answer consistent with a big bang event is that the light from the more distant stars has not reached us yet. The ones we see are close enough that their light has taken less than 13.8 billion years to reach us.

Bonus 2: Your Suntan

When you are on a beach working on your suntan, about 0.001 percent comes from photons originating from the big bang. Another 0.000000001 percent comes from stars not in our solar system. A full 77 percent of your skin darkens due to direct sunlight, and the remainder is triggered by light reflected off the sky.[26] This is known as the greenhouse effect, a huge topic of chapter 11.

Bonus 3: The Brief History of Our Sun

In the beginning, roughly 4.5 billion years ago, a cloud of interstellar hydrogen gas filled the universe. Hydrogen is the most abundant element in the known universe. As the cloud cooled (thanks to obeying the laws of thermodynamics, described in detail in chapter 21), it rotated and contracted due to gravity. As the cloud continued to condense, more and more pressure was applied to its hydrogen core.

The pressure caused its temperature to rise. The hydrogen nuclei were pushed together with such tremendous force that they fused, creating helium nuclei. This nuclear fusion produced a force that prevented the sun from collapsing further, and—presto!—there was light.

This is about where we are now in the sun's history. When the sun runs out of hydrogen to fuse, the outward energy will eventually fail. When this happens, our star will collapse once again, applying tremendous pressure to its helium core. This pressure will cause the helium nuclei to fuse into heavier elements until carbon is achieved. But all of that is in the future, the topic of chapter 21.

Bonus 4: The On and Off Lighting System of a Cepheid Star

Here is how astrophysicists use Cepheid stars to calculate distances. A Cepheid politely increases its temperature, size, and brightness at regular intervals that scientists can observe. Even more accommodating, the amount of brightness is proportional to its period. So if an astrophysicist knows how often it pulsates then, with a bit of math, she knows how bright it is. Now all she has to do is aim her telescope at the Cepheid and compare her results with its observed visual brightness. By comparing the two, she will discover how far away it is.

We can go deeper and ask why a Cepheid pulsates. It is caught in a feedback loop called the *Eddington valve.*

1. As the star compresses, it heats up. The helium in its outer layer becomes ionized. This is fancy talk that means the helium loses its electrons.
2. The star becomes more opaque, which dims the star. Temperature begins to increase, and the star becomes unstable.
3. The outer layer pushes out against the compression, and the star expands.
4. As the star expands, the helium becomes less ionized.
5. The less ionized the star is, the more transparent it becomes.
6. The brighter the star is, the cooler it becomes.
7. As it expands, gravity kicks in, forcing the star to contract again.

CHAPTER 5
PARALLEL WORLDS

Parallel worlds and parallel universes are staples of science fiction. In the original *Star Trek* series, I seem to remember an evil Kirk from an alternate dimension who had a pointy ear friend with a goatee.[1] Comic books are rife with parallel worlds. In both the Marvel and DC Comics universes, the mechanism of choice is quantum branching. DC Comics had their Crisis on Infinite Earths series, and Marvel had the Secret Wars story arc.

In the CW network's television series *The Flash*, how many parallel Earths has the hero Barry Allen been to? The television series *Sliders* had an endless slew of universes. The heroes were even confronted by Kromaggs, a race descended (probably) from a different evolutionary branch than Homo sapiens. The entire raison d'etre of the television series *Fringe* is the mysteries presented by a single parallel universe.

Plenty of examples of alternate universes pop up in movies. *Edge of Tomorrow* is based on a very enjoyable book with the scary title *All You Need Is Kill* written by Hiroshi Sakurazaka. The hero is stuck in a time loop and relives the previous day each time he dies in a war with alien invaders. Each redo changes the day, creating a parallel timeline from the one in which he died.

In books, try the Merchant Princes series by Charles Stross. The series is about a family with an inheritable trait (this trends to fantasy) that grants the ability to travel between parallel Earths. The family works as drug runners who do their businesses across Earths.[2]

I think you get the idea. Parallel worlds are a science fiction staple. But being popular in fiction does not mean they aren't real. This chapter covers different theories that can explain an alternate

Earth or an alternate version of you, the reader. They are all mathematically consistent and reasonably justified. Be warned, however, that none have been scientifically proven yet.

Brace yourself, because each one requires really, really large numbers.

PARALLEL WORLDS FROM MATH

The equations that describe the big bang have more than one solution. Each solution could be interpreted as another version of the universe. In fact, string theory has 10^{500} different solutions for the big bang.[3]

PARALLEL WORLDS FROM DISTANCE

We live in a large universe, a place larger than what we can observe. If we assume that the universe is infinitely spread out but not necessarily infinitely old (remember that the best estimate of its age is 13.8 billion years) then combinations of events will repeat.

Consider how unlikely your existence is. The odds against the combination of atoms coming together to make you—*you*, not a clone or a facsimile—are so fantastically small. The odds are probably less than 1 in $10^{2685000}$ lifetimes (a 10 followed by 2,685,000 zeros). Nevertheless, here you are reading this book. Congratulations on winning the life lottery!

Let's say you get a comic book every time you win a game of Jenga. It turns out that you reliably win about one out of every five games. If you played fifteen times, how many comics would you expect to win? The answer is three.

I'm not trying to insult you with simple math. I'm preparing you for the next paragraph.

Imagine that you played an infinite number of Jenga games and your average number of wins is a meager 1 in $10^{2685000}$. It sounds pathetic, but, given enough time, you will win. Given even more time, you will win again. (Don't go out and buy a lottery ticket because you suddenly feel lucky.) For this example, I assumed infinite time instead of infinite space, but I think you get the idea.

This theory uses probability to prove that, within an infinitely large (or near infinite) universe, other regions of space will be like ours and, no matter how unlikely, there is another planet like Earth and another version of you. And because we know some combination of atoms have created at least one Earth (the one you're living on right now), there has to be a nonzero chance for another one somewhere really, really far away.

Whatever else you can say about a 1 in $10^{2685000}$ chance, it isn't zero. So given enough space, there will be another you.

Something to ponder: if we add infinite time to the infinite space scenario then there not only exists another you somewhere contemporaneously (meaning in the now) in a huge universe, but it also implies that other you(s) have existed in the past and will exist in the future.

The distance argument, in general, makes a good case for the existence of nonhuman extraterrestrial life. If the universe is big enough, even given a low probability, other intelligence must evolve somewhere.

PARALLEL WORLDS FROM BRANCHING

This is the many-worlds interpretation of quantum mechanics described in chapter 2. I hope you recall the wave equation, where the size of a particle's wavelength dictates all the possibilities of the particle's position. And I hope that you remember that particles are both particle(ly) and wave(y).

The many-worlds interpretation considers a standing wave that holds up many branches. No, this is not a mixed metaphor. Any real possibility within the probability wave becomes a separate branch of the universe. The wave never collapses to a single outcome.

The Man Who Folded Himself, written by David Gerrold (he also wrote the fan-favorite *Star Trek* episode, "The Trouble with Tribbles"), is about some funky paradoxes caused when his character time travels to be with himself. Each time he travels, a new branch is created, and that new branch contains yet another him. Many versions of him, along with gender changes, appear in the story.[4]

Robert J. Sawyer's Neanderthal Parallax Trilogy (*Hominids*, *Humans*, and *Hybrids*) is set on a parallel Earth where the Neanderthals evolved and Homo sapiens went extinct.

In the movie *The Butterfly Effect*, a college student discovers he can make an alternate version of his present time by having his younger self make small changes in the past. (The actual term *butterfly effect* comes from chaos theory and will be covered in chapter 11.)

This movie is a lot like *Edge of Tomorrow* where the hero is trying to find the best future for himself (himselves, actually). The big difference between the movies is character placement. One lives in the present and manipulates his past self while the other lives in the present trying to affect the future.

MEM(BRANE) THEORY

Membrane theory is a derivative of string theory. This cosmology theorizes higher dimensions. The theory proposes that we live on a three-dimensional membrane situated within a wider multidimensional space. We (might) share this space with many different universes, and each universe can have very different physical laws, constants, and initial conditions.

Is life still possible in a universe with different laws of nature? This is an idea David Brin explored in his novel *The Practice Effect*. The protagonist travels to anomalous worlds in alternate universes where the physical laws are different.[5]

Can these different universes ever meet? String theory math shows that they can. But for us, let's hope not, at least for a very long time. When they do, it's a collision—a big bang. At least according to this part of string theory.

THE *WE LIVE IN THE BEST OF ALL WORLDS* THEORY

Forget the earth or the sun as the center of the universe. The anthropic principle puts life at the center. This is more philosophy than science because it cannot produce a falsifiable prediction or any testable experiment, but a lot of scientists have given it consideration. This is an idea I will not spend much time on. It is a lot of tail chasing.

The universe appears fine-tuned for life. If gravity were just slightly stronger, stars would compress more tightly and burn out after only a few million years rather than billions. Ergo, life would never get the chance to evolve. If the strong nuclear force were just a bit stronger, all the protons in the early universe would have paired off and water would not exist.

For this theory to come close to science, the anthropic principle relies heavily on the existence of a vast number of universes. This ties into string theory because some versions of string theory predict a multiverse wherein each "verse" formed with different constants. If we focus on regions of the multiverse where life forms, then we will (probably) find the constants predicted by string theory.

More on this circular thinking in a few paragraphs, but first we need to define the *weak* and *strong* versions of the anthropic principle.

> **Weak anthropic principle:** We live in a special place and time in the universe during which life exists. Surely the universe contains all the necessary parameters for life to exist because (guess what?) we are here.
> **Strong anthropic principle:** The laws of physics are biased toward life. The universe must be obliged to contain all the necessary parameters so that life can exist because (guess what?) we are here.

The difference between these definitions is nuanced, but (possibly) has philosophical existential consequences. The weak anthropic principle puts limits on certain properties of the universe. The fact that at least part of the universe contains carbon-based life observers puts constraints on what the whole universe can be like. For example, the universe must be at least old enough for evolution to have occurred. The strong anthropic principle implies that the universe is compelled to have properties compatible with intelligent life.

The unavoidable and uncomfortable part of this is the observation selection effect, also called *anthropic reasoning*. This is when the thing being studied is correlated with the observer. If humans hadn't evolved, humans wouldn't be around to study the probability of their evolution. Everything we observe is being observed by, you guessed it, us. To be observed, there must be an environment conducive for the observer. Tail chasing.

A nonreligious mechanism behind the anthropic principle (philosophy) might be Chaotic Inflation. This twist on the big bang redefines the cosmic inflation described in chapter 4. Inflation still occurs in different regions of the universe, just not necessarily all parts at the same time.

The standard theory considers inflation a one-time event. If Chaotic Inflation is true then different areas of space are undergoing inflation and evolving into separate universes. In turn, this chaotic inflation repeats itself in each new universe. Among this infinite number of universes, all different physical laws would exist. Through the pure odds allowed by an infinite number of universes, one of them must function under the laws that allow for stars, atoms, and life. Most of the rest would have different laws of physics and be barren.

PARTING COMMENTS

Parallel universes can be justified theoretically, but there is no practical evidence that they exist. It is all speculation, and no test that can prove their existence is anywhere in sight. In none of the proposed theories of parallel worlds—whether from math, distance, quantum mechanical branching theory, the bulk of string theory, or (dare I suggest) philosophy—is there any way for us to interact with anyone or anything in a different universe.

The idea is especially fun in science fiction. And, for our sanity's sake, it makes interpreting some of the physics of the universe easier (disclaimer: easier doesn't make it right). For example, the branching theory is a convenient solution to the grandfather paradox. This doesn't mean branching is real.

In science fiction, who cares?

CHAPTER 6
POWERING UP OUR CIVILIZATIONS

Science fiction that provides unlimited energy to a city, a space station, or a spacecraft is weak. Power is a limited resource and should be treated as such. No matter how advanced the civilization, a scramble for resources always ensues. Nations have risen and fallen over access to energy.

We need energy to survive. This is true for biological systems, plant life, nonbiological machinery, and, ultimately, the universe itself. Energy is the oomph we get from thermal sources such as the sun, chemical sources, mechanical motion, electricity, and nuclear reaction, to name a few.

Think about the time you finally decided to shovel your sidewalk after a blizzard. The energy for this project probably started out as chemical, meaning the Pop-Tart you had for breakfast. This chemical energy is transferred to the mechanical motion of your body shoveling the snow. If this were me, a portion of the food energy would also have been spent on swearing as my back began to ache.

Once humans created fire, we had the energy to stay warm and cook food. Eating cooked food uses less of the body's energy for digestion. More energy is available for the brain and, in the long run, allows more branches to be added to the evolutionary tree.

Speaking of trees, wood was one of our first nonfood power sources. Today, civilizations use combustible fuels such as oil and coal, or they use nuclear reactions. They also use renewable energy such as solar, wind, and water. And most recently, biofuel has been added to the mix.

CIVILIZATION RANKINGS

Now let's get to the fun stuff, rankings. I'm talking about civilization rankings. In 1964, astronomer Nikolai Kardashev created a scale to categorize how technologically advanced a civilization might be based on its energy usage.[1] Without further ado, here is the Kardashev scale:

A **type I** civilization is able to harness all the power available on a single planet. They have complete planetary control. Using Mother Earth for reference, a type I civilization can harness 10^{16} to 10^{17} watts. That is a one followed by sixteen or seventeen zeros. Our civilization is classified as type 0, at a little more than 70 percent of what it takes to be upgraded to type I. An example of a type I civilization in fiction is *Buck Rogers*.

A **type II** civilization is able to harness all the power available from a single star. Using the measured luminosity of our sun for reference, this comes to about 3.86×10^{26} watts. Examples in science fiction are all of the *Star Trek* mainline races (Federation planets, Klingons, etc.).

At the upper end of the type II range is the Dyson sphere. This megastructure is used a lot in science fiction but is based on science. It is named for mathematician Freeman Dyson who described how an artificial structure could completely encompass a sun and capture its power output.[2] The first science fiction description of such a structure was made by Olaf Stapledon in his 1937 novel *Star Maker*.[3] In it he describes "worlds constructed of a series of concentric spheres." In Larry Niven's novel *Ringworld* a Dyson sphere can be considered a main character. The description of how Ringworld can exist is hard science fiction at its best.

I hope you are ready for something really cool, something that ranks as cocktail conversation. About 1,500 light-years from here in the Cygnus constellation is a star called Tabby's Star (named after Tabetha S. Boyajian, its discoverer) that dims and brightens in odd but repeated patterns.[4]

A lot of speculation surrounds what this might mean. One unproven but fun explanation relevant to this chapter is that it could be a signal from an alien megastructure surrounding a star. The dips in light are too significant to be from a passing planet. A science fiction explanation is that it is a giant structure similar to the Dyson sphere.

A **type III** civilization is able to harness all the power available from a single galaxy. The measured luminosity of the Milky Way is approximately 1×10^{37} watts.[5] In fiction, such civilizations include the Borg from *Star Trek: The Next Generation*, Asimov's Foundation universe, and the Empire or First Order in the Star Wars franchise.

In the DC Comics universe, an example would be the Guardians of the Universe—the Green Lantern's bosses. But do not consider Marvel's *Guardians of the Galaxy* as an example. These guardians rank only at type II in the ability to use power.

Civilizations at types I through III constitute the original categories of the scale. Those classified greater than type III enter the realm of science fiction, so there isn't complete agreement on the divisions between the post–type III varieties. My presentation of them can be debated. Also, their energy usage is extrapolated.

A **type IV** civilization is able to harness all the power available from a supercluster. Our local supercluster includes the Milky Way galaxy, the Andromeda galaxy, and the forty-seven thousand much smaller galaxies in the Virgo Supercluster.[6] The power projection is 10^{42} watts. In fiction, type IV civilizations would be the Ancients in the *Stargate SG-1* universe and the First Ones (the Vorlons and the Shadows) in the *Babylon 5* universe.

A **type V** civilization is able to harness all the power available from the observable universe. It might not be possible for us to detect the existence of such a civilization because we are in the universe from which they are drawing energy. We could only perceive their energy usage as laws of physics. Their power usage is projected at 2×10^{49} watts.[7] These would be the Gallifreyans of *Doctor Who*.

A **type VI** civilization is able to harness all the power available from multiple universes. This type of civilization would have learned how to alter the laws of physics that apply to different universes. As a bonus, type VI can pack up and move when their universe dies. The death of a universe is the subject of chapter 21. Just for now understand this: it happens.

The power projection for a type VI civilization trends toward infinity. In fiction, I like writing about type VI civilizations when I can. In my short story "Chronology," published in *M-Brane SF*, I had a few lost citizens from a type VI civilization interact with us type 0 types. Mayhem ensued.

A **type VII** civilization is able to create a universe and then harness the power of each universe they create. These civilizations must remain outside the universes they create. This amounts to deity status. This might appear in fiction as mythology.

In the bonus materials for this chapter, I offer an alternative classification of civilizations. Instead of power usage, it considers size. As always in physics, size matters.

ARE THERE QUANTUM ENERGY SOURCES?

Yes, and they are based on two quantum phenomena that are exploited a lot in science fiction: virtual particles and zero-point energy. And guess what? They both owe their existence to our old friend from chapter 2, the Heisenberg uncertainty principle.

Let's make quantum mechanics weirder by bringing energy into our discussion. Because it is a function of wavelengths (the cause of all things fuzzy), its measurement has uncertainties. Like momentum and position, these uncertainties can't be reduced to zero simultaneously. These uncertainties give rise to virtual particles and zero-point energy, both of which have been used liberally in science fiction techno speak for energy.

Virtual particles are little somethings that are allowed to arise from nothing . . . as long as they promise to return back to nothing after a duration too quick to be observed.[8] These virtual particles permeate all of space, doing some very helpful things such as regulating particle decay and mediating the exchange of forces between particles.[9]

For example, when two negatively charged electrons repulse each other, they are exchanging virtual photons. These virtual particles are little messages saying, "Hey, you! Back off!" Because these virtual photons exist only for a short time, they can't travel very far, unlike lower-energy photons (let there be light).

POWERING UP OUR CIVILIZATIONS

This explains why the electric force is stronger at short distances. In fact, all the basic forces described in the first interlude diminish with distance for this reason. The caveat is that, although gravity also diminishes with distance, physicists have yet to reconcile this force with quantum mechanics.

Space is not as empty as you might think. Nature abhors a vacuum, so space seethes with virtual particles winking in and out of existence. Thanks to pesky particles popping up throughout the universe, action occurs at every point in space and time. Everything, everywhere, oscillates. Virtual particles are the quantum white noise of the cosmos. The energy from all that Jell-O-like quivering is called zero-point energy, which, by the way, is always *nonzero*.[10]

IS THERE EVIDENCE OF VIRTUAL PARTICLES?

The best-known experimental evidence of zero-point energy is called the *Casimir effect*. In 1948, Dutch physicist Hendrik Casimir predicted that a dense metal plate in a vacuum (the unlikeliest place to find energy) would be bombarded on both sides by virtual particles.[11] If you put two such plates very, very close, there won't be enough space between them for larger virtual particles to pop into existence.

Because the vacuum pressure is now less between the plates than on their outer surfaces, they experience a net force pushing them together. The effect was successfully tested in 1997 by Steve K. Lamoreaux of the Los Alamos National Laboratory.[12]

Another name for zero-point energy you might hear about in either science magazines or in science fiction is *vacuum energy*.

VIRTUAL PARTICLES AND ZERO-POINT ENERGY IN SCIENCE FICTION

Virtual particles are everywhere, and if they could be harnessed, imagine how advantageous that would be for colonization, war, and all the other ways we like to utilize cool things we discover.

Just as a caveat, remember that zero-point energy is already the lowest possible energy of a system. You have to be a savvy science fiction creator to come up with a plausible way of collecting it without using up more energy than you get out. According to physicists, extracting this energy is unlikely. Not so for the creative fiction writer!

A decrease in zero-point energy is known as *negative energy*. If a civilization could control this energy, a reduction in zero-point energy in front of a spacecraft would reduce resistance (negative energy would pull) and rapidly accelerate a spacecraft to near the speed of light. If this is a full type III civilization then perhaps with the aid of tachyon particles, the ship's acceleration might exceed the speed of light (temporarily exiting conventional space to do so).

Tachyons are hypothetical particles believed to have *never* traveled slower than the speed of light. The speed limit rule only applies to mass that started out slower than the speed of light. What is often forgotten is that in special relativity, the rule is symmetrical: anything traveling faster than the speed of light cannot travel slower than the speed of light.

PARTING COMMENTS

A lot of the time, science fiction provides unlimited energy for whatever civilization it describes. We know better. Fuel for energy is a limited resource, even when mining thermal energy from a black hole. Even capturing unlimited zero-point energy consumes limited resources.

The Kardashev scale is a ranking system of societies based on power usage. Do they rely solely on planetary resources? Do they use solar power? Do they mine black holes? How about batteries powered by virtual particles? Zero-point energy? No matter what, the Kardashev system has a ranking for them.

CHAPTER 6 BONUS MATERIALS

Bonus 1: An Inward Look: An Alternative Classification of Civilizations

Instead of energy usage, the Barrow scale classifies technological civilizations by their ability for inward manipulation, the control of smaller and smaller entities.[13] A lot of our -ologies, like biotechnology, nanotechnology, and even information technology, come from our ability to manipulate at small scales. Barrow believed there is more to explore in small scales than large ones. Plus there is no speed of light limit. The Barrow scale lays out the following:[14]

A **type I-minus** is capable of manipulating objects larger than the scale of themselves by mining, building structures, and joining and breaking solids.

A **type II-minus** is capable of manipulating genes and altering the development of living things, transplanting or replacing parts of themselves, and reading and engineering their genetic code.

A **type III-minus** is capable of manipulating molecules and molecular bonds to create new materials.

A **type IV-minus** is capable of manipulating individual atoms, creating nanotechnologies on the atomic scale, and creating complex forms of artificial life.

Our civilization is transitioning between Type III minus and Type IV minus.

A **type V-minus** is capable of manipulating the atomic nucleus and engineering the nucleons.

A **type VI-minus** is capable of manipulating the most elementary particles of matter (quarks and leptons) to create organized complexity among populations of elementary particles.

A **type Omega-minus** is capable of manipulating the basic structure of space and time.

Bonus 2: Comparing Energy Sources

In comparing various fuel sources, the energy returned on energy invested (EROEI) ratio is sometimes used. It shows how much energy is released relative to the amount of energy needed to get at the resource.

Resources with the highest EROEI are hydroelectric, coal, and oil. Although coal and oil originally held very high EROEI, the value is declining—it costs more and more to find and dig up these fossil fuels. The EROEI for finding oil went from 1,200 in 1919 to 5 in 2007.[15] This means we get five times more energy from fuel than the energy expended to find it. That is still a lot, but it is declining.

The problem with the EROEI metric is that it does not explicitly account for environmental costs. These costs necessarily increase as the difficulty of acquiring fossil fuels increases. Because of falling EROEI, alternatives such as wind, natural gas, solar, and nuclear are being considered more and more.

The sun provides clean energy, but how do we exploit it? We currently use photovoltaic panels made of silicon, but solar panel efficiency is nothing to write home about. The record in 2014 was 46 percent efficiency.[16] The typical solar panel is about 20 percent efficient.[17]

Besides panels, the sun's energy can be captured by letting plants do the work for us. Biofuels such as ethanol (from corn) is made from seeds. With this method, however, the hazard is that farmers will plant seeds for fuel instead of food. This can be especially problematic in developing countries.

CHAPTER 7
BLACK HOLES SUCK

It's a pity that nobody has found an exploding black hole. If they had, I would have won a Nobel Prize.

—Stephen Hawking

Brace yourself. Parts of this chapter will give you flashbacks to all the general relativity goodness you read about in chapter 1. Einstein's equations of general relativity demonstrate the positive relationship between mass and gravity. The more massive an object, the more spacetime dips around it. The bigger the dip, the greater the gravitational field. A black hole is the mother of all dips.

Black holes come in different sizes—they can be as small as an atom with the mass of a mountain, or enormous with the mass of over one million suns (called a supermassive black hole).[1] It is believed that every large galaxy contains one of these supermassive black holes at its center. As discussed in chapter 1, the supermassive one at the center of our galaxy, the Milky Way, is called Sagittarius A*. The size of central black holes might play a role in how galaxies form.

HOW DO BLACK HOLES ARISE?

No definitive answer for this exists, but we do have one strong possibility based on a star's lifecycle. The power of a star comes from the nuclear reaction taking place in its core. The reaction causes outward pressure that is balanced by the weight of gravity from the star's mass. The nuclear power pushes out while gravity pushes in.

As the star gets older, it changes its process from combining hydrogen into helium to fusing helium into carbon. Then it changes carbon into oxygen, and then oxygen into silicon, and finally silicon into iron—a cosmic dead end. Iron is a stable element, so the energy output from fusion is no longer possible. Without fusion producing outward pressure to challenge gravity, the outer layers of hydrogen, helium, carbon, and silicon that were previously created burn around the iron core.

At this point, the size of the star plays a big role in its destiny. Stars below what is called the Chandrasekhar limit, roughly 1.4 times the mass of our sun, will collapse into a white dwarf.[2] Their cores stopped fusing at carbon because they don't have enough mass to generate the gravitational pressure necessary to overcome the electromagnetic repulsion between protons. This is why there is no fusion of carbon into the heavier elements.

With bigger stars, those above the Chandrasekhar limit, the iron core continues to build up as these successive layers burn through their remaining fuel and then crash down on the core. The core is compressed further until it completely loses the battle with gravity and collapses, causing a colossal explosion. Materials, including the heavier elements beyond iron, are blasted outward. This is called going supernova.

If the original star had a lot of mass, say ten times that of our sun, then more fireworks arise after the supernova explosion. The core compresses further, ramming atoms against each other and transforming protons into neutrons until—voila!—we have a neutron star.

If the star was over twenty-five times the size of our sun, then the neutrons can't prevent the gravitational pressure from collapsing the star further. The star will occupy a smaller and smaller portion of space while maintaining its mass until it becomes an infinitely small, dense point known as a gravitational singularity—the heart of a black hole.

This is where the known laws of physics break down. A gravitational singularity presses so deeply into the fabric of spacetime that it creates a gravity well steep enough that the escape velocity is greater than the speed of light. No escaping light means it goes dark.

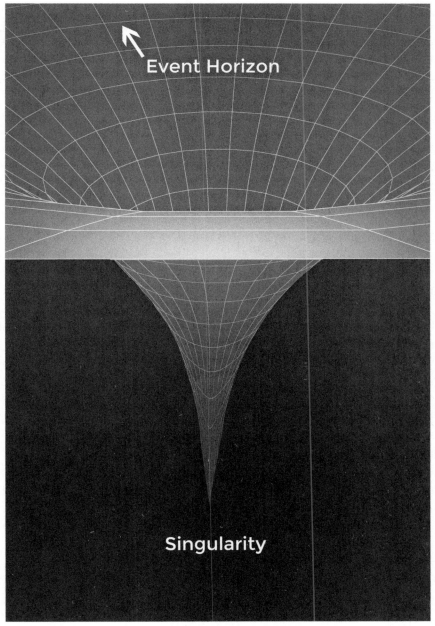

Fig. 7.1. Illustration of a singularity.
(Modified from image by iStock Photo/Yurkoman.)

Named for Karl Schwarzschild, who solved Einstein's general relativity equation for a star's gravitational field, the Schwarzschild radius defines the radius of a given mass at the point at which no force could stop it from continuing its collapse to a singularity. When an object is compressed smaller than its Schwarzschild radius, it becomes a black hole.

The estimated Schwarzschild radius of the earth is nine millimeters. If the earth could be compressed down to that size, it would collapse into a black hole.[3] Our sun would have to be compressed to a radius of three kilometers before it turned dark and holey. Don't worry, this won't happen to either the earth or the sun. Probably.

WHY DOES ESCAPE VELOCITY MATTER?

Escape velocity is an important concept if you plan to launch a rocket from Earth or anything else with significant mass. If you throw a ball up into the air with all your strength, even if you have an arm strength of NFL quarterback Aaron Rodgers, it will slow, stop, and fall back to Earth. Gravity works against your best efforts. To escape Earth's orbit, you need to heave the ball at escape velocity, the minimum velocity needed to escape a massive body.

It shouldn't come as a surprise that the more gravity (mass) a planet has, the greater the escape velocity becomes. You would need to toss that ball at about eleven kilometers per second (km/sec) or twenty-five thousand miles per hour (mph) to escape Earth. Now, if you are on Jupiter, you would need to hurl the ball at 59.5 km/sec (133,018 mph).[4] The escape velocity of the sun is 618 km/sec (1,381,601 mph). The good news is that if you plan to chuck a rock off the moon, you need to accelerate it to a scant 2.4 km/sec (5,369 mph).

The escape velocity of a black hole is where things get interesting, at least for a cosmologist. Maybe not so much for the engineer and pilot aboard a starship, however. The escape velocity is greater than the speed of light (300,000 km/sec).[5] If your starship flew *somewhat* near a black hole then its path will begin to curve toward it. A good pilot can probably avoid it.

If you pass too close then no pilot can prevent your ship from falling in. The critical distance between escaping and falling in is called the *event horizon*. Think of an event horizon as a fence with a big Keep Away sign hammered into it. Personally, I would do what it says.

HOW DO BLACK HOLES RELATE TO SPAGHETTI?

If you do decide to ignore the warning and trespass on the event horizon, I hope you like pasta. This is not an adventure I would recommend. However, if you insist, the first thing you do is put on the latest spacesuit and disembark from your starship. There is no reason to endanger the rest of the crew.

In space, you orientate yourself toward the black hole feetfirst so that you can watch your approach to the event horizon. According to general relativity, as you enter, you feel a strong gravitational pull on your feet. This pull will get stronger and stronger relative to the amount of gravity pulling down on your head. You begin to stretch. This continues until you are stretched into something resembling a skinny noodle, referred to us humans as spaghetti. Scientists call this effect *spaghettification*, and I bet it isn't fun.

If you prefer to consider this thought experiment from a quantum point of view, then you will simply burn up when you hit the event horizon. An event horizon that follows quantum rules will be highly energetic. I bet you still wouldn't have fun.

DO BLACK HOLES LAST FOREVER?

No, black holes do not last forever. They emit small amounts of radiation and evaporate away. In 1974, Stephen Hawking showed that black holes evaporate via Hawking radiation.[6] He discovered something with his name. What were the odds of that?

Anyway, Hawking radiation arises when pairs of negative and positive virtual particles pop into existence near an event horizon. Normally they would annihilate each other, but because of their proximity to the black hole, they are unable to complete their marriage contract and are forced to part ways. One of them gets caught in the event horizon while the other escapes into space.

The particle that falls into the black hole has negative energy, thereby reducing the mass of the black hole, thus creating evaporation. Recall that, according to general relativity, mass and energy are equivalent. When the black hole sucks down a negative energy particle, it is sucking in negative mass. And as with most things in physics, size matters. The smaller the black hole, the quicker it loses mass. As it shrinks, its temperature will rise until it explodes. Or it might simply fade away, depending on your theory of choice.

On a technical note, the above is just a mental visualization of math. As the potential energy evaporates from the black hole (positive energy) it must be offset by the shrinking potential energy stored in the black hole to conserve the overall amount of energy in the universe. This subtraction is why it is called negative energy. It is a math thing.

An evaporating black hole could be a good thing. It is shedding energy back into the universe, so in principle it could be harnessed as an energy source. Of course, a much higher-level civilization, probably a type III, would be required to mine all the thermal (hot) energy. How could that happen, you'd like to know?

To make it worth the hassle, they would have to stake their claim on a large black hole. Because, you guessed it, size matters. A lot of time is required for the large ones to evaporate enough energy to make harvesting the energy worthwhile, something along the line of about 10^{57} x 13.8 billion years. The lifetime of a black hole is the cube of its mass (m^3).[7] And these large suckers have a lot of mass.

Fear not. When a civilization approaches the late stage of type I status, the idea of having a lifespan is quaint. Time might be on their side for these types of energy projects.

HOW ARE BLACK HOLES DETECTED?

Black holes exist mathematically, but science demands evidence. They can't be seen optically (here it comes) because they are black and so is space. Their existence and size can still be detected by indirect means. You can use the known mass of objects you can directly detect to estimate the mass of a black hole by how it affects those known stellar objects.

Another method is to measure the X-rays emitted by heated material as it falls into the black hole. A third way comes from another prediction of general relativity called *gravitational lensing*, the amount that background light is warped by massive objects. The more massive the stellar object, the more the light is curved.

WORMHOLES (AGAIN)

Let's pair up what we know about quantum mechanics with our knowledge of black holes. Only instead of entangled particles, I want you to think of entangled black holes. The quantum state of one black hole is in sync with the quantum state of a second black hole. If instruments measure the quantum state of the first black hole, guess what? You know about the second one because everything that is entangled on the quantum level is in the same state.

Now consider the possibility that a singularity isn't a point but really is a hole, a wormhole that connects two black holes. Each black hole resides in a different place and perhaps even in different times. The black holes are physically connected, and the entanglement causes a wormhole because of entanglement's effect on the geometry of spacetime.

Recently, more and more physicists have been looking at the effects of quantum entanglement on spacetime. A few even believe that quantum entanglement itself might create spacetime.[8]

CAN BLACK HOLES BE USED FOR TIME TRAVEL?

In principle, yes. However, it means stretching the physics of general relativity to the limit, which is okay at least for good science fiction. Here are a few ways it could be done.

106   BLOCKBUSTER SCIENCE

1. Go for a spin

Fig. 7.2. Illustration of a spinning black hole. (NASA/JPL-Caltech.)

The first way requires a rotating black hole. According to general relativity, a spinning black hole twists the surrounding fabric of spacetime. This phenomena is known as *frame-dragging*. I hope you still have that two-dimensional sheet we used in chapter 1 pulled taut with a big ball lying on it. Now spin the ball. It pulls the sheet along for the ride.

The same holds true for objects spinning in our three-spatial-dimensional spacetime sheet. Scientists have tested this for the earth using gyroscopes aboard NASA's Gravity Probe B satellite.[9] The frame drag of forty-two milliarcseconds of angle over the course of a year is consistent with the prediction from general relativity. This isn't a big number, but space around Earth is dragged.

The technology doesn't yet exist for a satellite to test a black hole for twisting spacetime. That doesn't stop scientists from thinking up more practical ways to conduct that test. They came up with the idea to study stuff circling the ingeniously named H1743-322 black hole through the X-rays emitted from iron ions embedded in that circling matter.[10] The X-rays grow or shrink depending on how the ions move relative to an observer. The frame-dragging effect is about one hundred trillion times as strong as the one found near Earth.

Now all you need to do is have your pilot fly your ship into the spin cycle as everyone holds on. You and your crew have just entered a different time frame.

2. Care for a donut?

Another possibility for time travel that uses a rotating black hole is a ring singularity (think donut). This type of singularity is called a Kerr singularity, named after mathematician Roy Kerr who solved Einstein's field equations of general relativity for a rotating black hole.[11]

And now we mix some science fiction with the Kerr singularity. You start by creating a starship with enough density to survive the gravity gradient and maintain its shape as it passes through the event horizon. Step one completed: spaghettification avoided. Now all that is left is to dive through the center of the ring and emerge in a different time.

3. Time dilation (general relativity) because sometimes you just have to go with the classics

Another way to time travel is to camp out in a starship near the edge of the event horizon and take advantage of the time dilation. Remember that as gravity increases, time flows slower for those in the starship relative to their companions on a settlement very far away. An example of this can be found in the television series *Andromeda* where a starship uses this method to travel three hundred years into the future.

PARTING COMMENTS

Black holes are where the physics of both quantum mechanics and general relativity break down. Black holes are gravity wells pressing so deeply into the fabric of spacetime that light and electromagnetism can't climb out. They exist and hold galaxies together, but they can't be completely defined by general relativity or quantum mechanics. They don't last forever. For the record, black holes don't actually suck you in. Instead, you fall into them.

CHAPTER 7 BONUS MATERIALS

Bonus 1: A Massive Mystery

Supermassive black holes are believed to start out small. They gradually grow by taking in matter and by merging with other black holes. Astronomers have detected the existence of these big guys in the early years of the universe, back when it was a youngster of less than a billion years old.

The mystery is that although the first stars were huge (one hundred times the mass of our sun) and burned for only a few million years, their explosions should have created black holes of about the same mass as the original star. And yet, some of the supermassive black holes weigh in at least ten billion solar masses.[12]

Bonus 2: The Information Paradox

It is generally accepted that, if you have information about a system such as particle states and their quantum probabilities, you should be able at any time to determine its state at any other time. In other words, the quantum probability wave that describes the states must be conserved. This conservation holds in quantum mechanics and general relativity until a black hole is introduced.

As we've seen, these critters have the habit of gobbling up everything in their paths, including quantum information. That means the quantum wave is incomplete. Back in chapter 2, I described how a probability wave holds all possible states of a particle until it is observed. At that moment, the wave collapses and its state can be measured. All the future possibilities of a particle were included in the wave function.

After a black hole has destroyed at least part of that wave, we can no longer know everything about the particle. This contradicts general relativity, which has time dilation preserving the system information. This contradiction is known as the *information paradox*. A theoretical solution will be provided in chapter 20. It questions what we know of reality.

CHAPTER 8
ORIGIN AND EVOLUTION OF LIFE ON EARTH

Evolution is cleverer than you are.

—Leslie Orgel, evolutionary biologist

This chapter is dedicated to the opportunism that has built life on Earth. Whether the cause was natural selection, random mutations, selfish genes, dark energy, or chemical reactions doesn't matter. Things have changed over time, and the things we see now didn't exist in the past. The reason is evolution.

Evolution affects everything in the macro universe from life to stars to black holes. As long as time has a direction, evolution occurs. There is no escaping it. For organic life on Earth, evolution is the backbone on which life sciences are built. It is the name given to the slow process (in human terms) of inheritable changes in populations over time.

Evolution can be confusing and even controversial because it is sometimes referred to as fact and other times as theory. It is both. A fact describes a phenomenon while a theory attempts to explain it. This is no different than our discussion on gravity. You can see that gravity exists (a fact) by simply dropping this book. The theoretical explanation for gravity is the curvature of spacetime. This theory might not be complete because we know it doesn't jibe perfectly with quantum mechanics. Nevertheless, gravity exists.

Evolution is no different. The *fact* of evolution is backed up by a lot of evidence such as DNA mapping and the fossil record. Evolution *theory* is used to explain the *how*. Charles Darwin believed the how of evolution to be natural selection, where a variation that provides a survival advantage in a population is passed down through the generations.

His theory was predicated on the Malthusian principle of population growth. Reverend Thomas Robert Malthus argued in his book *An Essay on the Principle of Population* that populations grew exponentially, but the ability to feed the increasing population grew at a lower geometric rate.[1] The human population, he claimed, grows at a faster rate than its food supply.

Darwin had much the same idea about species. Since they produce more offspring than the number that can be supported in a given environment, some members of the population will be better suited to survive in that environment and more likely to mate.

With natural selection, those members more likely to survive are more likely to reproduce. Natural selection also weeds out individuals with unfavorable traits, called *survival of the fittest*. Survival of the fittest does not mean survival of the smartest, or that higher intelligence is the outcome of evolution. All that matters is the passing on of genes. In Kurt Vonnegut's novel *Galapagos*, humans in a distant future have evolved into sea creatures that laugh at farts.[2]

How these variations arise in a population was a mystery to Darwin. Today we know they are caused by random changes in DNA. More specifically, evolution is a function of genetic mutation, a mistake (alteration) in a DNA sequence created when a cell copies itself for cell division. Occasionally the mistake gives the organism a survival advantage.

Surviving (obviously) enhances the likelihood of meeting a mate and reproducing, which passes the mutation to the next generation. As I said at the beginning of this chapter, it is all about opportunism, selfish genes passing themselves down to future generations. This same process eliminates bad mutations from the gene pool.

Some people confuse adaptation with evolution. This is easy to do because they are related. A member of a species who adapts to a hostile environment is more likely to find a mate than members who weren't able to adapt . . . and died. Adaption is a short-term phenomenon where the survivor is still a member of the original species; the species itself hasn't changed. Evolution is a much longer process where physical changes begin at the genetic level and take generations to produce species more suited to an environment. Adaption is about the short-run survival of an individual member while evolution is the long-run survival of the entire species.

Evolution to the rescue: Between 1996 and 2016, about 80 percent of Tasmanian devils were wiped out from a contagious cancer called devil facial tumor disease (DFTD). The small number of survivors had a genetic variant that helped them survive the disease long enough to reproduce. They evolved themselves out of extinction.[3]

THE HUMAN STORY

You are both a hominin and a hominid. Congratulations. These terms differ in a subtle way that can cause confusion. A hominin is any subspecies of early humans more closely related to humans than chimpanzees. Hominids include all hominins plus all modern and extinct apes, gorillas, chimps, and orangutans. You are also human, part of the homo genus. In fact, your species is the last of its kind.

Our story reveals how evolution is a steady process but not necessarily a straight-line progression. Darwin's book, *On the Origin of Species by Means of Natural Selection, or the Preservation of Favoured Races in the Struggle for Life*, expressed his belief in branching evolution from a common ancestor. He was correct. There are many branches in our family tree. Most were pruned away because they couldn't adapt to changing environmental conditions.

Here is a flash history of how we got from Earth's start to hominins traipsing about to humans messing about on Twitter:

- 4.5 billion years ago (BYA), Earth forms.[4]
- 2 BYA, oxygen on Earth increases significantly.
- 650 million years ago (MYA), multicellular life begins in the oceans.
- 440 MYA, life climbs onto land.
- 250 MYA, dinosaurs appear.
- 65 MYA, dinosaurs go extinct.

- 4 to 3 MYA, *Australopithecus afarensis* stood on its hind legs and walked. This hominin is the common ancestor to both the genus *Australopithecus* and the genus *Homo*. You might have heard about the *A. afarensis* celebrity known as Lucy.[5]
- 2.5 to 3 MYA, the climate changes, forcing *Australopithecus* to move out of the forests and into open territory. They adapted to new food and new predators. The genus *Homo* begins.

Something to ponder: if the climate had not changed then, *Australopithecus* might still be around today.

- 3 to 1.5 MYA, *Homo naledi* buried their dead. *Homo habilis* mastered stone technology. *H. naledi* were *Australopithecus* sized but had some of the modern *Homo* features (bigger skull size). They might be the earliest of *Homo* species.
- 1.9 MYA, a chinless *Homo erectus* started a fire.

    The movie *One Million Years BC* (originally released in 1940 with the famous remake produced in 1966) shows how dinosaurs and humans chased each other around. Do you want to know if this is fiction? I'll tell you.

    Both the historical and fossil records tell quite a different story about what was really going on at this time. In one million BCE, *H. erectus* were picking up stones and definitely did not look anything like Loana the Fair One of the Shell tribe, as played by Raquel Welch. Also, there weren't any dinosaurs.

    Perhaps the title contains a typo and the writers of the movie really meant 100,000 BCE when humans were working their way through the Stone Age. The problem, again, is that there weren't any dinosaurs during that time either, so I suspect *One Million Years BC* wasn't produced in a documentarian style.

    The movie *10,000 BC* (2008) is a little closer to accuracy, but somehow the Nile Valley of 2,000 BCE makes an appearance along with ice-age Egyptians who wield steel and ride domesticated horses.[6] So I'm guessing this movie didn't rely much on the historic record.
- 780,000 to 125,000 years ago in Africa, *Homo heidelbergensis*

split into at least two groups. One stayed in Africa while the other migrated to parts of Europe and Asia. Individuals who hunkered down in Europe evolved into *Homo neanderthalensis* (aka Neanderthals), while those who planted their flag in Asia evolved into Denisovans (status pending, but probably in the *Homo* genus). The branch of *H. heidelbergensis* that stayed in Africa eventually evolved into Homo sapiens.

- 75,000 to 50,000 years ago, the largest migration of *H. sapiens* left Africa.

    There is a theory that our species almost did not make it beyond this point. About seventy-four thousand years ago, a volcano in Indonesia caused a volcanic winter that lasted six to ten years.[7] This gray weather was followed by a thousand-year-long cooling. This might have reduced the human population to ten thousand to thirty thousand individuals. It is called the Toba catastrophe. The good news is we haven't gone extinct, at least not yet. That is a topic for another chapter.

    Now back to our history. The largest migration of humans out of Africa happened between fifty thousand to seventy-five thousand years ago, possibly because of climate changes. All non-Africans alive today descended from this group. There is early evidence of humans hiking about, but we didn't descend from them. Those early migration humans all died out. Scientists aren't sure what happened. Perhaps the massive migration overwhelmed the smaller earlier migrations physically and genetically. Whatever happened, only bits of their DNA made it down to us today.

- 25,000 to 40,000 years ago, *H. neanderthalensis* dies out.
- 10,000 to 20,000 years ago, the *Homo floresiensis* branch dies out and Homo sapiens *sapiens* is the last species of the genus *Homo* standing. That's you.

    H. floresiensis gets the dubious nickname of "hobbit" from J. R. R. Tolkien's diminutive protagonist. These historical hobbits were about half the size of *H. sapiens*. They were hanging out on the island of Flores in Southeast Asia when modern humans wandered across Europe. Did humans meet them and start an oral story tradition about dwarfs? Possibly.

Anyway, this branch of the family tree is a mystery. One theory is that they are a descendant of some early *H. erectus* that settled on the island about one million years ago. They then evolved smaller bodies to accommodate limited island resources. In evolutionary terms, this is tricky because they would have had to shrink in the span of a mere 300,000 years.

FOSSILS ARE SO YESTERDAY (LITERALLY)

Scientists no longer need only fossils to reconstruct human evolution history. Today they also rely on genetics. During the Stone Age, all these different hominins conducted a lot of fraternizing. The DNA results are in, and there is no hiding paternity (and maternity). Between 1.5 percent and 4 percent of Neanderthal DNA can be found in today's non-African genomes. Between 1.9 percent and 3.4 percent of modern Melanesians genes come from the Denisovans.[8]

There is also some genetic evidence that Neanderthals and Denisovans might also have thrown some adult private dating parties. In addition, there is evidence that Neanderthals and Denisovans mated with other archaic humans about 120,000 years ago. Apparently they couldn't wait for modern humans to come along.

WHOA! SLOW DOWN AND KEEP IT REAL

In science fiction, evolution into our future tends to occur much quicker than it did in our history. In the classic 1968 movie *Planet of the Apes*, somehow by 3978 CE intelligent apes rule devolved humans. The premise is fun, but the science is shaky. If you take the difference between 3978 and 1971 (the year Charlton Heston's character George Taylor's rocket first launched), you get 2007 years. Not much time for traditional evolution.

As long as we are looking at a two-thousand-year window for drastic evolutionary changes, I should mention the Wayward Pines trilogy. In the books, aberrations (called abbies) evolved from baseline humans in less than two thousand years.[9] The next time you hear

ORIGIN AND EVOLUTION OF LIFE ON EARTH   115

about the next phase in human evolution (think X-Men), be suspicious. It's probably just funky genetics (comic book meta-genes?) causing stuff and not natural selection.

HOW DID LIFE BEGIN ON EARTH?

Obviously, before life can evolve, it must exist. All preexisting life comes from earlier preexisting life, except when it does not. At some point, abiogenesis, meaning life from nonliving matter, occurred. Both philosophically and scientifically, this is a pretty big deal.

Let's start at the beginning. Not the beginning of everything, but I'm talking about the origin of life on Earth. For clarity, I will use NASA's definition of life: a self-sustaining chemical system capable of Darwinian evolution.[10] The chemical reaction part refers to the metabolic process, or obtaining energy from organic sources, that supports life.

Life most likely began on Earth about four billion years ago. During the Hadean eon, Earth was a young and rambunctious five-hundred-million-year-old, the sun wasn't as bright, the moon was closer, and Earth spun faster. It spun so fast that a day lasted a mere ten hours.[11] Oxygen wasn't present because there weren't any flora to produce photosynthesis waste. Meteorites pummeled the planet, providing it with some of our metals.

Against this chaotic backdrop, scientists might be able to trace back to the origin of earthly life. The theory is still murky, so the tale I'll tell is one of scientific possibility. Obviously this will be a rough draft because we are all waiting for biologists and chemists to confirm these events and further flesh out details.

Somewhere in Earth's history, chemicals (probably simple chemical polymers) developed a handful of biological properties such as self-replication and cooperation with other molecules.[12] This is the era of chemical evolution when simple chemicals began making copies of themselves. Over billions of generations, more complex variations emerged.

Eventually (let's say about 3.5 billion years ago), some variations were enclosed in a membrane, possibly from a fatlike substance that surrounded the molecules with a bubble. The first biological chemicals might have occurred after millions of years of chemical reactions assisted by ultraviolet light, lightning strikes, or deep-sea vents. These became the first microbial cells from which life evolved—abiogenesis. Each generation of these protocells produced new versions of themselves until eventually they split into bacteria and archaea.

Quick fact: The you who is reading this isn't completely human. Your body contains about the same number of microbial cells as human ones.[13] And because of evolution, that's okay. They coevolved with us.

About 2.4 billion years ago, things got interesting. Simple organisms called *cyanobacteria* (a phylum of bacteria) evolved the ability to convert sunlight into energy, and photosynthesis was born. During this era, the buildup of oxygen led to the ozone that later would protect life from the sun's ultraviolet rays.

Members from each family (bacteria and archaea) started dating and gave birth to more complex organisms. At some point, an archaea cell was taken hostage by a bacteria cell. Instead of killing the archaea, the bacteria enslaved it within itself. This was the first eukaryotic cell. A eukaryote cell contains a nucleus and other goodies like mitochondria, a power plant that provides energy to the cell and allows for complex organisms to arise.

Fast forward to about six hundred million years ago when multicellular life began to flourish. Boom: the Cambrian Explosion.[14] During the next thirty million years, complex organisms popped up all over the place. Most of the major animal phyla appeared. The Cambrian period corresponds directly to the rise in oxygen levels that was probably caused by algae and moss fixing carbon in the soil.

WHAT IS DNA?

Cells contain self-replicating material—deoxyribonucleic acid, aka, DNA. This molecule is composed of two coiling strands of phosphates and sugars that form a double helix. It contains the blueprint for the proteins and molecules used by human cells for growth and reproduction.

ORIGIN AND EVOLUTION OF LIFE ON EARTH 117

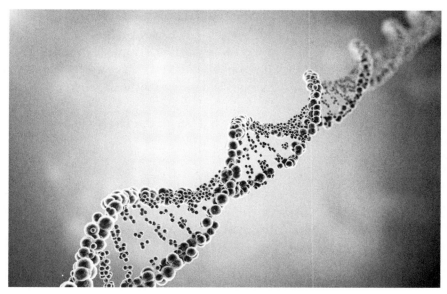

Fig. 8.1. Illustration of DNA. (iStock Photo/Rost-9D.)

Nearly every cell in one person's body has the same DNA. If you've watched even one episode of *CSI*, you know that one person's DNA is different from everybody else's DNA. Almost. My mother and father's DNA is the most like mine, unless I have an identical twin. I don't, but if I did he would be the exception. Our DNA would be the same.

Quick fact: Not all "sames" are created equal. Although twins might have the same genome, it might still be possible to tell twins apart by their DNA, especially if they have been exposed to different environments or are living different lifestyles. These differences cause epigenetic differences between their DNA. Researchers have found that different epigenetics lead to different melting points for the DNA.[15] This can be used for identification. See bonus 2 to learn more about the genome and epigenome.

We can go deeper and ask, how did DNA originate? It is *extremely unlikely* to have been formed by a chemical accident. In order to make DNA, you need specific enzymes (proteins that cause reactions). To make those enzymes, you need precise instructions carried within DNA.

Guess what? You are trapped in a circular conundrum similar to the classic question over whether the chicken or the egg came first. By the way, the answer is the egg. Whatever laid it wasn't a chicken, but it was genetically close.

Here are two leading theories that break the circle on how DNA evolved.

1. Homegrown DNA

    DNA wasn't the first chemical to come to life. A lot of biologists believe the honor belongs to ribonucleic acid (RNA), the half-dressed version of DNA. Yes, this risqué molecule wears only a single biopolymer strand.

    RNA is like a construction worker who carries out the instructions in the DNA blueprint. On a planet of opportunities such as ours, it should be possible for construction workers to also be their own blueprint. Scientists have shown that RNA can spontaneously fold itself into structures that form self-replicating molecules. Therefore, in theory, RNA could act as both as information carrier and a catalyst (both worker and blueprint). It is very possible that RNA first evolved from the chemistry of early Earth. At some point, RNA life transitioned into the more complex DNA life we see today.

2. Imported DNA

    Another idea is that life might not have originated on Earth but rather was seeded with ancient microbes that bombarded the planet billions of years ago. This idea is called *Panspermia*, the Greek word for "seeds everywhere." If true, this would mean that the chemical evolution happened elsewhere and spread to Earth by piggybacking on the backs of asteroids or by swishing around in the water-ice of comets.

    A cool title for a book might be *Ribose from Outer Space*. And there is a possibility it would be nonfiction. Ribose (of RNA fame) might form on dusty ice that had been irradiated by ultraviolet radiation. Comets are made of ice, and they have been known to strike the earth. So make of that what you will.

ORIGIN AND EVOLUTION OF LIFE ON EARTH

The seeding of biological matter doesn't have to be at the interstellar level. We could look closer to home. Our neighbor Mars has water. In the deep past, it might even have hosted oceans where organic matter could have formed. The early Earth lacked some of the elements necessary for life such as boron, which is needed to form RNA. While boron was rare on early Earth, it was less so on Mars. Meteorites from the Red Planet might have come loaded with the element. It might not be science fiction that Martian RNA led to Earth DNA.

I know what you are thinking. The seeding could have been a bit more intentional, something planned by an alien civilization. Science fiction often agrees. In the 1993 *Star Trek: The Next Generation* episode called "The Chase," we learn that the Star Trek galaxy was seeded by ancient aliens with DNA codes similar to the characters' life-forms.[16]

Thus the Star Trek galaxy is full of two-legged, two-armed, two-eyed races that, except for a little face makeup, look very similar. This also might explain why humans and Vulcans are able to interbreed. There is a distinct lack of speciation in the Star Trek universe. Speciation is the evolutionary process by which isolated populations from the same originator species eventually become so distinct that if they bump into each other again, they will no longer be able to breed.

42

—Douglas Adams

Let's not forget Douglas Adams's wonderful *The Hitchhiker's Guide to the Galaxy*. The earth was built as a computer program, seeded with human life, and supervised by mice. Why? To determine the *question* to life, the universe, and everything . . . the one that might or might not go with the *answer* to life, the universe, and everything.

PARTING COMMENTS

Life began from those chemicals that developed biological properties like self-replication and cooperation with other molecules. These protocells branched into bacteria and archaea. Then one day, along came the eukaryote cell with its fancy nucleus, mitochondria, and DNA. From then on, life could not be stopped.

To get to us it took a bit more evolution. Evolution is the slow process of inheritable changes of populations over time. Natural selection is an explanation of how these changes are passed down to successive generations. The variations are caused by random changes in DNA. Okay, cue up the ominous music because natural selection decides which mutations live and which die in the gene pool.

Scientists no longer use fossils alone to reconstruct human evolutionary history. They now have DNA mapping. They have learned that hominins had many evolutionary branches. Almost all of those extras were cut off because the subspecies couldn't adapt to changing environmental conditions or competition from other subspecies. In the end, only one type of hominin was left: Homo sapiens *sapiens.*

CHAPTER 8 BONUS MATERIALS

Bonus 1: Evolution Is Smart

Evolution doesn't guarantee that every mutation will be a long-run improvement. But a mutation that doesn't necessarily look so great at first might prove to be pretty good. Take our inability to make vitamin C (ascorbic acid). This vitamin is necessary to produce collagen and to prevent connective tissues and blood vessels from breaking down. Almost all animals and plants produce their own vitamin C. Why not us?

An intriguing theory is that our ancestors had a vitamin C–rich diet, so it was more efficient to eat it than to produce it. This redirected our bioenergy to building a more resourceful (and bigger) brain. An alternate theory is that our ancestors' vitamin C deficiency damaged their DNA. Mutation rates increased, and our evolution sped up.

Bonus 2: The "Ome" Home

Let's take a ride up the taxonomy elevator of your genetics. The first floor is the genome represented by DNA. This is where your genes come from. The next floor up is the transcriptome level where you are greeted by RNA, which is busy switching genes on and off. Up another floor, you reach the proteome level where you can say hi to the proteins.

Shall we continue up? Good. The metabolome level is waiting to show you all the small molecules in the cells. Up again to discover the epigenome level where the environment plays a role in gene expression. Finally you arrive at your destination, the phenome level, where your traits (physical and behavioral) are determined from the activities on all the floors beneath.

The genome is the most stable over time. The other levels are a bit more flighty and switch genes on or off at particular moments. The genome describes what *can* happen, while the proteome and the metabolome reveal what is *actually* happening. Two people with the same genome can have very different phenomes.

There are ways to influence the way genes function without changing the underlying DNA sequence. Stress, poverty, and pollution can accelerate aging. These genetic expressions occur at the epigenome level and are called epigenetic traits.

SECOND INTERLUDE
MORE OR LESS HUMAN

> *I teach you the overman. Man is something that is to be overcome. What have you done to overcome him?*
>
> —Friedrich Nietzsche, *Thus Spake Zarathustra*

Is aging a disease? Should extending life and increasing happiness through biological hacking or by harnessing technology be a goal? Is there any reason you can't choose your gender on demand? Is there even a reason for gender?

Someday (soon?) we will be able to make ourselves smarter, stronger, and more attractive. Improved people are another staple of science fiction. But predicting what humans will self-evolve into is tricky (and fun). We might enact biological improvements down to the DNA level, or we might blend our bodies with machine (cyborgs), or become completely synthetic, or do away with bodies entirely and become enhanced uploads.

The road to improved people is often bumpy, and no speed limit has been posted. This truth of science is exploited in good science fiction. For example, consider a fictional Earth where at some time in its (future) history, it will be popular to experiment with memory enhancing drugs and antiaging treatments. Plausible.

Now, I trust the readers of science fiction to know that small changes become cumulative over time, so allow me to continue. As more time passes, perhaps my fictional population begins to add nanotechnology (technology nearly at the atomic level) and bioengineering to the mix. How about neural interfaces to connect their brains to information webs? After all these changes, are they still human?

What I described above is called transhumanism—the human in transition. It is a stage in human development where human beings use technology rather than evolution to improve the race. Genetic modification, gender reassignment, upgrading with prosthetics, implants, and/or various chemical enhancements are all on offer. We already conduct biological hacking by replacing knees and hips. Why not cheat death by replacing aging or damaged organs with ones grown from the body's cells?

If these cumulative changes keep occurring, the mental and physical capabilities of augmented beings will at some point so radically exceed those of baseline humans that they might no longer be considered human. Posthumanism (when the body becomes irrelevant) might come after that. In fact, posthuman thinking and experiences might be so profoundly different from ours that we can't conceive it. Posthumans will shape themselves as well as their environment.

A lot of arguments blaze about modern ideas to improve humanity. The following chapters reveal that evolution doesn't have to be random. Humans can take over the human condition with science and technology. Chapter 9 covers how we can use technology to modify our bodies on the road to transhumanism. Chapter 10 describes how we might add technology directly to our bodies, merging us with machine.

Death is an existential threat. It is the end of consciousness, and yet we cannot consciously imagine not being conscious. All we have to do to bypass this paradox is to live forever. Simple. Using technology externally and internally is a way to stall death until we can download our minds into a virtual (posthuman) world housed in an indestructible data storage device floating safely in deep space.

Concerns about the sun dying out or Earth being overtaken by hostile aliens become moot. You or your progeny will be able to exist indefinitely, or at least until you read the bad news in chapter 21 (how everything ends).

CHAPTER 9
BADASS BIOLOGY

> *How I, then a young girl, came to think of, and to dilate upon, so very hideous an idea?*
>
> —Mary Shelley

In Mary Shelley's story, Dr. Victor Frankenstein is so wracked by grief he buries himself in an experiment to give something nonliving a life. He creates a large humanoid using repurposed parts (from corpses). Eventually, he abandons his creation and mayhem ensues.

This chapter is less about creating life and more about hacking it.

GENETIC ENGINEERING AND EVOLUTION

Passing down favorable traits using good, old classical evolution (via breeding) is so yesterday. I think all we hominins can agree that natural selection/random mutation evolution has had a good four-million-year run, but, as Bob Dylan might say, times are a-changin'.

We no longer need this low-tech type of genetic engineering. Today we can go a lot less natural with our selections. We can identify specific genes and insert them directly into the DNA of an embryo, passing down a trait in a single generation. A trait not already mainstreamed in the population.

To avoid future disease, doctors have been tinkering with the human genome through gene therapy for some time now. They also use gene-editing technology to discover which DNA sequences are present in diseases. Humans can be edited. Instead of just removing negative traits, this same technology can be used to engineer more positive ones. The question is, should we? The answer for a lot of people is yes. If enough people agree then we might be past the age of random mutation and into the era of intelligent human design.

Here is something outside of science to consider: is it ethical to impose eye color, intelligence level, or sexual orientation of *your* choice on the next generation?

HOW IS GENETIC MODIFICATION DONE?

In the field of biology, humans now can "cut and paste" using the newest and hottest gene editor called the CRISPR/Cas9.[1] CRISPR stands for clustered regularly interspaced short palindromic repeats. The CRISPR can be used to fight viruses, eliminate genetic diseases, prevent HIV infection in human cells, and for the contentious idea of creating designer babies.

The CRISPR uses guide RNA to lead the enzyme Cas9 to a targeted section of DNA for a little nip and tuck. The RNA is chemically paired with the DNA so it can hone in on its goal. Once on target, the Cas9 acts like scissors to snip problem areas of DNA. This process can eliminate genetic diseases such as breast cancer, cystic fibrosis, sickle-cell disease, Tay-Sachs disease, and many more. And as I mentioned before, this device might also be used to insert something into the cut, providing specific advantages to a baby.

WHAT DOES SCIENCE FICTION SAY ABOUT THE ETHICS OF GENE MODIFICATION?

> *The right of a child to be born to its full potential shall take precedence over all other considerations. It is the state's responsibility to safeguard the legal rights of the intended embryo.*
>
> —Thirty-Ninth Amendment to the Constitution ratified July 4, 2051, from Bruce T. Holmes's *Anvil of the Heart*

In *Anvil of the Heart* by Bruce T. Holmes, parents who do not optimize the genes of their children do not find themselves in a comfortable situation.

Science fiction is great for examining ethical perils. The 1997 film *Gattaca*, written and directed by Andrew Niccol, is about gene manipulation beyond disease prevention. The manipulation ensures that children inherit the best traits from their parents. The depth of this movie comes from Vincent Freeman, the main character, overcoming—or rather, escaping—genetic discrimination because he is born the old-fashioned (classical) way.

Quick fact: the word "Gattaca" comes from the first letters of the four nucleobases (biological compounds) of DNA: guanine, adenine, thymine, and cytosine.

Tired? Did you have a long night? Wouldn't it be nice to never need sleep or feel fatigued again? You could be much more productive and successful at your profession. Rodger Camden hopes exactly that for his daughter in Nancy Kress's novel *Beggars in Spain*. He arranges for her to be genetically modified in utero to grow up never needing sleep.[2]

Good idea? Maybe. In her story, the trait isn't without problems. The sleepless crowd is discriminated against because it is believed that they have unfair advantages in business. This makes sense. Employers could hire one employee who can work sixteen hours a day instead of two ordinary humans. Populist movements support buying services and products only from normal sleepers. Now imagine the mob's reaction when they learn that a possible added benefit of being sleepless is immortality.

SPEAKING OF IMMORTALITY, IS IT POSSIBLE?

> *I don't want to achieve immortality through my work: I want to achieve immortality through not dying.*
>
> —Woody Allen

The human body carries genes handed down to our children. Our bodies evolved to be disposable, to make room for future generations. New technology has changed all this. Can aging be abolished? Someday answers, "Possibly." Science fiction answers, "Definitely."

Plenty of evidence proves that the average human lifespan has increased with healthier diets, better housing, and better medicine. And as discussed above, gene therapy to an embryo still in the womb (or inside a science fiction womb substitute) can have its DNA optimized to extend life by eliminating potential disease.

As they say, "it's all in your genes." And whoever *they* are, they are correct. But of course you already knew that from the chapter on evolution. In a Stanford University study of people who have lived at least one hundred years, scientists have uncovered 281 genetic markers that slowed their aging and made this group less susceptible to disease.[3]

Scientists are learning so much about genes (and still have so much more to learn). The cerebellum ages slower than any other part of the body. The cerebellum is the region of the brain responsible for motor control and cognitive functions. It appears to stop aging around the eighty-year benchmark.[4] So if you happen to be one hundred years old, your cerebellum would have been immune to deterioration for the previous twenty years. If researchers can find the genes responsible for this, who knows? Maybe you live forever.

Or we could borrow some advantages from other species. Sea life would be a good start. A Greenland shark is the current record holder for the longest lifespan. Based on carbon found in the shark's eye lenses, one shark was estimated to be 392 years old.[5] The bowhead whale can clock up to an impressive two hundred years. As weird as it sounds, the DNA of different species can be combined. Think of the genetically modified glow-in-the-dark cats on YouTube. So maybe we can incorporate this whale trait into our own DNA.

> *There is nothing in biology yet found that indicates the inevitability of death.*
>
> —Richard Feynman

It is in the cells of your body that your genes hang out. It is good that they have a home (they built it), but your cells are another source of aging. They have a limited shelf life (the notable exception being cancer). They can divide until they hit between forty and sixty doublings. This is known as the Hayflick limit, named for Leonard Hayflick. He discovered that, in addition to external wear and tear, the limit on cell division is part of what makes us die.[6]

It turns out that cells have a memory of sorts. A meter inside the nucleus reminds a normal cell how many divisions it has already gone through. If a group of cells were frozen and thawed out a few weeks later, they would continue dividing right where they left off until they hit their Hayflick limit.

The limit originates from the shortening of the DNA telomeres with each new division. Telomeres are the sheaths at the end of chromosomes. Think of how the aglet at the end of a shoelace frays with wear. As we grow older, our telomeres dwindle, and the "fraying" impairs the immune function. Telomere shorting is a biomarker of aging.

A protein called Gata4 acts like a switch that forces cells to stop growing and dividing.[7] As we age, we accumulate more and more of this protein. It might someday be possible to reverse this switch in older humans. This is not for children. It should only be attempted after tissues are finished developing; otherwise, organs won't develop correctly.

Worn-out tissue can also be reconstructed using stem cells. Stem cells are undifferentiated cells capable of differentiating into specialized cells. They have the potential to become any type of cell such as blood cells, skin cells, brain cells, etc. Using stem cells, biologists have successfully grown an organoid that imitates the human midbrain, the part of the brain that regulates hearing, vision, and movement. This organoid can be used to test new medications for Parkinson's disease and other brain diseases.[8]

IS A LONGER LIFE WORTH IT?

This is up to you. Here are some questions to get you thinking:

- Is a longer life expensive? If yes, will the poor revolt?
- If older (in terms of age and not body) people don't retire, how do young people get jobs?
- If funding is diverted to support the long-lived, what happens to funding for the education of the young?

How would this affect marriage? Can people tolerate each other for eighty to ninety years? Will serial marriage become the norm?

CAN'T GET ENOUGH OF YOURSELF? SEND IN THE CLONES

If your body parts wear down, they can be replaced. Someday it might be common to have replacement organs grown from your own cells. Why stop there? Why not clone your entire body and harvest the organs as needed, or use it as a younger blood donor? It would do wonders for your energy level and recuperative abilities.

Cloning is asexual reproduction from a single ancestor. The clone will be genetically identical to its progenitor; i.e., a cloned human is genetically identical to her parent. If done on a large scale, variation can only occur in the population via mutations.

Cloning today is done using the SCNT method. This stands for somatic cell nuclear transfer. First, the nucleus of an undifferentiated embryonic cell, which contains all of the donor's genetic goodies, is removed. The nucleus is then injected into an egg cell, and a small electrical shock is delivered to start cell division. The embryo is then implanted into a surrogate female and carried to term.

As with a lot of subjects in this chapter, ethical questions arise about cloning. In the future (if not in science fiction), will growing organs become a business? Should clones be exploited for organ transplant? How about creating a young clone to receive an upload of your mind?

What does individuality mean to a clone? Think of the Philip K. Dick classic, *Do Androids Dream of Electric Sheep?* This story was adapted into the movie *Blade Runner*. A lot of time in science fiction, cloning is about identity and slavery.

In the 2009 movie *Moon*, cloning is a form of slavery. Sam Bell has the lonely job of mining helium-3 on the moon. After three years, when his contract has concluded, he is sent back home to his wife and child.[9]

Only this never happens. Every three years a new Sam, a clone with the same memories of home and a wife, is woken to begin a three-year contract. The same again, and again. The movie *Oblivion* (2013) has a similar theme with Jack Harper clones tricked into drone repair on a dying Earth.[10] In both movies, there comes a moment when the clones question their identity. Identity is the path to freedom.

The clones in the Star Wars universe have no doubts about who and what they are. They were all grown to become cannon fodder, except the one who goes into bounty hunting.

GENETICS FOR THE ZOMBIE APOCALYPSE

The common cold was the (classical) biological weapon that saved our earthly rear ends in *War of the Worlds* by H. G. Wells. For us terrestrials, the cold virus is the least of our problems. Today we have weaponized smallpox, anthrax, Ebola, and rice blast (crop disease caused by fungus). You don't need science fiction to imagine military researchers tinkering with the genetics of these organisms to create more powerful weapons.

Now I'm going to cover something possibly worse, the zombie apocalypse. Zombies *might* make (fictional) scientific sense. In *Patient Zero* by Johnathan Maberry, people are infected (zombified) by the Seif al Din prion disease. In *World War Z* by Max Brooks, they are infected with the Solanum virus. In both cases, it is spread by biting.

Zombies exist in nature. Not human ones, at least not yet. For now, I'm talking about zombie ants. Brace yourselves for a horror story. The villain in this story is a fungus spore called *Ophiocordyceps*. The spore clamps onto an ant that probably was minding its own business. It digs its way into the ant's body and starts growing in the ant's head near the brain.[11]

At some point, about half the cells in the ant's head will have become *Ophiocordyceps*. This is when the fungus excretes a mix of chemicals that give it control over its host. The ant is zombified. The zombie is driven to climb a tree and clamp its jaws onto a leaf where it then proceeds to die. A stalk of the fungus sprouts out the body and drops new spores to the ground where they wait for unsuspecting ants so they can repeat the zombie cycle.

If I were forced to come up with a way to create a human zombie, I might try to genetically modify the rabies virus. Rabies infects the central nervous system and drives people to be violent. Now all an evil scientist (not me) would need to do is hybridize rabies with a flu virus to create an airborne contagious disease.

BACTERIA AND VIRUSES TO THE RESCUE

To fight off the zombies or other sundry villains, you might need special capabilities. A lot of super-powered characters pop up in comic books and movies. They appear to be stronger, faster, and more difficult to damage than a simple baseline human such as myself. And unlike the particle accelerator explosions and lightning storms that linked Barry Allen to the speed force in the CW television series *The Flash*, solid biological science can be used to create a superhero or supervillain.

Bacteriophages, or phages for short, are viruses (sometimes human-made) that seek out and infect bacteria, leaving human cells untouched. They penetrate bacterium and then go into a nonstop multiplying frenzy until the phages burst out. Think of the alien creature bursting out of officer Kane's chest in the movie *Alien*. Like the victims of the movie, when the phages are done with the bacterium they have invaded, they move on to the next bacterium. You could design a hero immune to specific bacteria strains.

Another type of phage is the macrophage. It is formed from monocytes, one of the groups of white blood cells that surround and digest pathogens and fix damaged tissue. This is a healing factor. Perhaps not Wolverine of the X-Men level, but it isn't too shabby.

Some of nature's greatest nanotechnology tools are viruses. Viruses are equipped with a few proteins and DNA strands, not too much stuff, and yet they are capable of sneaking into host cells and multiplying. With a bit of direction, they are used in gene therapy to deliver a healthy gene to a defective version of the same gene.

Instead of being an ally, some (actually, quite a few) viruses cause problems. For these cases, it is possible to use bacteria as a weapon. As we saw in the evolution chapter, a bacterium is capable of engulfing a virus and storing its DNA sequence. From that point on, if the same virus attacks, all future generations of the bacteria use the memory of the viral DNA to aim killer enzymes at it. This is similar to how scientists can disable genes or insert DNA sequences into an organism.

PARTING COMMENTS

Today's techniques of gene manipulation have begun to remove the randomness of evolution. Humans are taking over the human condition. Both lifespans and happiness can be increased with biological tweaks.

Clones of a human body can be used for organ harvesting or as a vessel into which to download memories. Or in science fictions, as slaves. Badass biology comes with a log of ethical baggage.

CHAPTER 9 BONUS MATERIALS

Bonus 1: Crime-Busting Science

How about taking some real science and mixing it with science fiction and mystery? Wouldn't it be cool if, without using a police database, a detective could determine the age and gender of a suspect from DNA left at the scene of a crime?

Forensic technology isn't quite there yet, but it's getting close. Researchers at the University at Albany found a chemical biomarker that can be used to estimate age has recently been identified in blood.[12] The enzyme reaches its peak level at eighteen years for a female and ten years for a male, and then declines with age. The same research team also found biomarkers that can be used to determine gender.

The same blood might also reveal a personality trait. Researchers have found certain genetic differences between early birds and night owls.[13] DNA left behind might reveal whether the person is a morning or evening type.

Now let's try some crime reconstruction. We've seen how blood can help with immediate profiling; now let's add how it might someday help reconstruct the crime. I'm going to go fictional on you and tell you about space detective Clyde Rex. He is assigned to solve a murder on the space-liner *Baxalor*. The first thing Clyde does is to kick everyone out of the space-liner's luxury cabin because he works alone.

He pulls out a laser scanner loaded with the latest imaging software that recreates blood spatter trajectories. Based on the size of the stain, his equipment can calculate the mass of each blood droplet. Next, he runs an algorithm on his computer pad that traces backward along the paths of every drop. If the crime scene has been cleaned to hide evidence, Clyde could use an infrared camera to detect blood proteins. Not all of this is fiction.

Bonus 2: Metamorphosis

Quick Fact: the DNA of any individual from any species will have the same DNA over its entire life.

So if the DNA of an insect doesn't change over its lifespan, what explains metamorphosis? Metamorphosis is a process through which a mature body becomes different than its adolescent form—unrecognizably different. The answer is hormones.

Hormones are chemical messengers that affect growth and metabolism. They also signal basic needs such as hunger and reproductive urges. For insects in particular, they trigger metamorphosis and encourage a larva to become an adult. In bugs, the hormone ecdysone changes how the creature's DNA expresses itself.

So the DNA remains constant, but hormonal influences change which genes are turned on and to what extent. This drives metamorphosis: cells produce different proteins that slowly cause a physical change. Consider a caterpillar. It chows down on some leaves until it reaches a critical size, at which point ecdysone is released. For the caterpillar, eating is the environmental trigger for ecdysone. Once again, nature has something to say.

CHAPTER 10
WELCOME TO TECH U

Man is an artifact designed for space travel. He is not designed to remain in his present biologic state any more than a tadpole is designed to remain a tadpole.

—William S. Burroughs

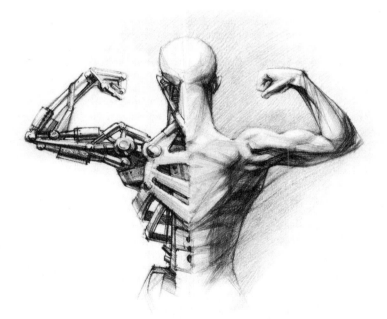

Fig. 10.1. Illustration of a transhuman (cyborg). (iStock Photo/Auris.)

Transhumanism is not only about how we change ourselves biologically, but it is also about how we can integrate our flesh with technology. It can be as simple as an artificial eye that allows its owner to see farther than baseline perception and, when required, to see in the dark. If we go a bit science fiction-y, our hero Hector could pop out his artificial eye and leave it behind. The eye sends encrypted signals directly to his brain, showing him who is following him. If he doesn't like what he sees, he can use his bionic arm to restrain the fiend until the authorities arrive.

Those are illustrations of the big stuff. Don't forget about the small. Nanotechnology can be used not only for repair but to enhance, to make you much better than you were originally. It can increase strength and memory. Or perhaps it really is all about looks. Someday we might be able to alter our features to impress a date or to go undercover as a spy. Say good-bye to plastic surgery. Best of all, none of this is impossible.

A human who has some bodily functions aided by or dependent on mechanics is called a cyborg. This is becoming more and more common in our everyday world. Surgical procedures today are at the point where replacing a knee or hip is considered (important but) mundane. Using advanced technology, doctors can address many types of physical disabilities. Robotic replacements for amputees are being developed that can be controlled using tiny electrodes implanted in the recipient's brain. The electrodes receive signals from the prosthesis, and the patient is able to feel and move fingers.[1]

In 2016, Zurich, Switzerland, hosted the first cyber Olympics called the Cybathlon. Disabled participants from all over the world competed using the most advanced prostheses.[2] It was a testing ground to determine how competitors coped performing typical daily tasks using the latest technologies. The events included bicycle races powered by brainpower and competitive bread cutting.

Science (as well as science fiction) needs to address the question of how accumulated improvements will change societies. If society gets in the habit of replacing its organics because of easier upkeep and regular upgrades, then you could end up with the Cybermen in *Doctor Who* or, yes, you got it, the Borg from *Star Trek: The Next Generation*. From these examples, I hope you see the potential for cyborgs as soldiers.

And video games are crazy with cyborgs. They typically can be found throwing cars around or converting their arms into laser cannons to prevent alien invaders from stealing children. Consider the characters in *Halo, Call of Duty: Black Ops 3,* and *Half-Life 2.* And let's not forget about comic book superheroes like the aptly named half-human, half-machine hero called Cyborg in the DC Comics universe.

Ghost in the Shell by Masamune Shirow started out as manga before being adapted for anime (the word *anime* comes from the pronunciation of the Japanese abbreviation for animation) and ultimately into a live-action movie. Manga is a Japanese style of comic book writing and cartooning. It is created for all ages, so for adults the themes can sometimes get dark or sexual.

Ghost in the Shell is about cyborgs with brain implants that can interface with technology. The only problem is that the brains can be hacked and all their memories replaced with false ones. Who can you trust when your perception of reality can be changed on demand?

CELEBRITY FICTIONAL CYBORGS

- Steve Austin (the bionic man from *The Six Million Dollar Man*)
- Jaime Sommers (the bionic woman from *The Bionic Woman*)
- Alex Murphy (law enforcer in the Robocop series of movies)
- Darth Vader and Luke Skywalker (a random father and son in the Star Wars franchise)
- Victor "Vic" Stone (from DC Comics)
- Nathan Summers (aka Cable from Marvel Comics)
- Metallo (the man with a kryptonite heart who battles Superman in DC Comics)
- Otto Octavius (aka Doctor Octopus, the man of many arms who wants to give Spider-Man a hug in Marvel Comics)
- Cyberman and Daleks (*Doctor Who* baddies)
- The Borg (living the life in a box in *Star Trek: The Next Generation* and *Star Trek: Voyager*)
- Tony Stark (the man without a heart; he sometimes dresses up as Iron Man, but the artificial heart makes him a cyborg)

THE TRANSHUMAN BRAIN

The human brain evolved to manage about eighty years of memories. What happens if genetically modified humans are able to live up to two hundred years? Ask any high school student, and if he is honest (of course he is), he just might confess that sometimes it is a struggle to remember recently learned concepts. How about you? Have you ever forgotten where you put your house keys?

Recent research seems to show that these memories are not lost but rather are stored somewhere that isn't easily accessed. Scientists at MIT are studying early-stage Alzheimer's disease in mice to show how memories aren't necessarily lost, but the ability to retrieve them might be faulty. Using light to manipulate neurons (a technique called optogenetics), researchers were able to bring back memories of being shocked to mice suffering from the disease. This shows that by directly activating memory cells, a memory can be recalled.[3]

And now comes the time for some tech speculation. How about simply adding a brain implant to store our memories instead of trying to manipulate neurons? In William Gibson's "Johnny Mnemonic" (*Omni* magazine, 1981), Johnny's implant system could hold hundreds of megabytes of data (considered a lot back in the 1980s). For this to be practical for patients with dementia, researchers would first have to find a way to transfer stored memories into active memories.

Why stop with memories? Perhaps someday brain chips could be used to improve cognitive function and reverse the aging process. Or instead of only internal improvements, a chip could enhance your external experiences. Someday it might be possible for you to control various devices in your house or, even better, use Wi-Fi to connect with someone else's chip for communication.

If we could do this now, then someone who traveled in a time machine from H. G. Wells's Victorian England to the present might think this looks like telepathy. This example gives me the perfect opportunity to invoke Arthur C. Clarke's third rule of prediction that "any sufficiently advanced technology is indistinguishable from magic."[4] Telepathy is magic. Soon, we might all be magicians.

SOMETIMES YOU NEED TO GO SMALL

Nanotechnology is technology manipulation at the nanoscale (one to one hundred nanometers). A nanometer is a millionth of a millimeter or a billionth of a meter. This technology can be used to manipulate molecular structures. So instead of the big add-ons like an arm blaster, you could fill up with the smallest of nonbiological tech, the nanite. These nanomachines are so small they measure less than one hundred nanometers (one-thousandth the thickness of a dollar bill).

Once fully developed, these little guys might be used in nanomedicine. They could be programmed to seek and destroy cancer cells, or they could hang around inside your body until needed to fix damaged tissue. Nanotech could ferry drugs directly to tumors, magnetically pull toxins from patients, replicate human organs for drug testing, or break up blood clots.

As long as I'm compiling a wish list, how about nanites that drive bacteria as a vehicle to transport molecules to damaged areas? Or use them as tiny missiles to destroy waxy plaque in blood vessels? All of this is possible. Fiction is not required.

A cool proof of concept for a nano drug delivery system already exists. Researchers have developed nanoparticle balls made from a polymer that disintegrates when ultraviolet light is shone on it.[5] The idea is that they can be filled with medicine, injected into the bloodstream, and, when the nanoball reaches its target in the body, light is applied and the medicine released.

In the movie *Fantastic Voyage*, based on a story by Otto Clemet and Jerome Bixby, a group of doctors are miniaturized and sent into a body to conduct brain surgery. Fiction aside, nano-submarines are real and can be powered by motors that operate much like a bacteria's flagellum (the tail-like filament that allows microbes to swim through fluids).

When excited by ultraviolet light, the bond that holds the rotor changes states, allowing it to rotate a quarter turn. As it returns to its resting state, it jumps again, rotating another quarter turn. This process continues as long as the light is on. It can achieve speeds up to 2.5 centimeters per second. It can carry cargoes for medical and other purposes.[6]

Nanites are the go-to panacea for some science fiction civilizations. The Borg automatons in the Star Trek universe have them flowing through their blood making constant adjustments. In "The Empty Child/The Doctor Dances" episodes of *Doctor Who*, we learn that Captain Jack Harkness's medical supplies include restorative nanites. And as you might expect, the Doctor happens to remember all about them just in time to save the day.[7]

Known as replicators in the television series *Stargate: Atlantis*, nanites were a little less helpful after they achieved sentience. It turns out that they weren't particularly fond of humans.

AND IN THE END, POSTHUMANISM

If we follow the path of augmentation, there will come a time when we have upgraded ourselves beyond the point where we might be considered human. I guess this isn't much different than how we stand currently as post-*Australopithecus*. Only this time instead of natural selection, our mastery of DNA hacking and technological augment is sitting in the driver's seat of evolution.

Many science fiction authors have written works set in a posthuman future. Let's not forget about the classic H. G. Wells novel *The Time Machine*. I hope the "post" we become isn't Eloi or Morlock.

Anyway, posthumans might not reshape only themselves but also their environment. In *Glasshouse* by Charles Stross, minds are uploaded into various bodies. Sometimes not on a voluntary basis, and sometimes not into the gender of their choice, and sometimes not placed in the right virtual time period.[8] I'll let you read the book to figure out what I mean by that.

A universe cohabited by humans and posthumans would have interesting dynamics. In Dan Simmons's novels *Illium* and *Olympos*, the posthumans dominate and are worshipped as gods.[9] In my short story "Chronology," published in *M-Brane SF*, they are godlike but are hardly worshipped. Then there is Hannu Rajaniemi's Jean le Flambeur book series, in which posthumans are a different species from baseline humans, and groups of posthumans are different from each other. Some of them actually get along. The others? I'm not telling.

How about this for a story? A contagious virus leaves 1 percent of its victims awake and aware but unable to move. This is the plot used by John Scalzi in his novel *Lock In*. As you might have guessed from the title, these poor people are locked into unresponsive bodies. Thanks to a government program, these victims are provided with robotic transports into which their consciousness can be transferred. They gain mobility while caretakers watch over their bodies.[10]

There are also people who can act as integrators, those who can let a locked-in person borrow their bodies for sensual experiences. Finally, a shared cyberspace-like environment allows the locked-ins to meet socially. Some choose never to leave that space. This series provides a good blend of virtual reality, transhumanism, and posthumanism.

In *Star Trek* the original television series, a repeating theme is that transcending the human condition leads to disaster (usually for the humans left behind). In season one, "Where No Man Has Gone Before" and "Charlie X," some humans turn posthuman, causing conflict. You also meet aliens who have transcended their original condition such as Trelane from "The Squire of Gothos."[11]

So much posthuman plotting in a single season! Don't even think about getting me started on the Q from *Star Trek: The Next Generation*.

CYBERING (EVERYONE IS DOING IT)

Although it would be unnecessary, especially when you are able to create your own virtual universe, there might come a time when your posthuman self wants to get social with other human abstracts. A good meeting place is a communal area called cyberspace.

The word *cyberspace* is credited to author William Gibson, who first used it in his short story "Burning Chrome."[12] The term doesn't make the leap from science fiction to the mainstream until after his novel *Neuromancer*, where Gibson describes cyberspace as a hangout where posthumans might live until the end of the universe.

Although much more mundane (for now), cyberspace is our

internet community. We can hang out there socially without the oh-so-retro demands of meeting a person in the physical world.

PARTING COMMENTS

Machine efficiency is greater than biological evolution. Transhumanism is not just biological fixes; it is also the upgrading of humans with technology. Posthumanism is when the accumulation of transhuman upgrades has changed our bodies and minds so profoundly that they might no longer be considered bodies and minds.

Something to ponder: After we have downloaded our minds into a collective intelligence, or into servers that generate virtual realities, we could make copies of our original baseline selves and watch them muddle around Earth for entertainment. Are you a copy?

CHAPTER 10 BONUS MATERIALS

Bonus: Nano for Food

Nanotech could make those "born on" dates or "best by" dates obsolete. Someday stores will be able to ensure safer and healthier products at your grocers. Nanoparticles could store the nutrients inside food and release them at a fixed time and fixed place in your body.

Conductive polymers molded into nano-sized sensors could detect the beginnings of molecular spoilage or foodborne pathogens, including those of nefarious (think bio-warfare) origin. Sticking with the theme, we should probably use degradable polymers to create the biosensors for food packaging.[13]

This technology exists, but consumers might encounter problems, such as death. Materials at the nanoscale might be harmful if they remain in the body too long. Most of what is being tested today is made of synthetic carbon materials. What if researchers used materials made of natural ingredients that break down in the body?

Someday sensors might help users analyze their breath to discover food preferences. After all, our breath chemicals correlate with our taste preferences. These same types of sensors could be used by doctors to check to determine whether dietary requirements or restrictions are being met.

CHAPTER 11
MAN AND NATURE

This distant image of our tiny world ... underscores our responsibility to deal more kindly with one another, and to preserve and cherish the pale blue dot, the only home we've ever known.

—Carl Sagan

Your home is more than the dwelling in which you huddle on a cold winter day while streaming *Star Wars: Episode IV—A New Hope* and sipping hot cocoa. The entire earth is your home. We are so perfectly adapted to this piece of real estate it would be very hard to live anywhere else in the universe. The earth has changed our species through our need to evolve in order to survive its climate changes.

A fascinating, and dangerous, detail is that the interaction is not one-way. Humans have also changed the environment. It is possible that sometime in the future, these human-made environmental changes might lead to our extinction.

As we evolved, our relationship with the environment due to our ever-increasing demand for energy has also changed. For a time, humans were content burning wood for heat, boiling water for steam, and burning calories for labor. Our modern lifestyles demand energy from many more sources. We have renewable ones like wind, water, and solar, but for the most part our current level of technology (proto–type I civilization) is much more suited for burning fossils fuels such as coal and gas.

Fossil fuels are combustible geologic materials created in the deep geological past from previously living matter (but not dinosaurs; that is a myth). Today we have so many people using up so much energy that we are endangering the environment on which we depend. This chapter focuses on environmental changes due to climate changes caused by our power usage. It offers up a few possible scientific fixes to save our collective butts. Some of them will sound like science fiction.

CAN CLIMATE OR WEATHER BE PREDICTED?

The earth's climate has never been constant. The three main drivers are energy from the sun, volcanic eruptions, and gasses trapped in the atmosphere. All three yield their greatest climate impact through their effects on the atmosphere. Remember how the volcanic winter of 1816 led to the story of Frankenstein's monster? Or the volcanic winter seventy-five thousand years ago that almost eliminated the human species? It's all caused by how the sun's energy reacts to what is spewed into the atmosphere.

Climate is the statistics of weather. Climate is therefore predictable because it is based on averages. We know it is colder in winter than summer. Unfortunately, weather isn't so easily predicted. In this chaotic system, small differences from an initial state can eventually cause large differences in the system. This is the so-called butterfly effect.

For example, hurricanes get their energy from the moisture of water evaporated off of tropical oceans. Land storms start with temperature differentials. Evaporation and temperature differentials are only two of the many variables used in constructing meteorological predictive models. Because of the chaotic nature of the model's variables, the further out the forecast reaches, the lower the odds are that the model's predictions will be correct.

That said, weather predictions are pretty good week out. After that, they are marginal at best. For long-term forecasts, the only really useful arrow in a meteorologist's quiver is her understanding of the connection between the atmosphere and the ocean. Tracking El Niño and La Niña, for example, correlates with weather patterns months ahead. Meteorologists can start to build a long-term forecast by tracking what these systems are doing.

Quick fact: Oceans have a higher heat capacity than the atmosphere because they are more stable. This means it takes more energy to heat the ocean one degree than to heat the atmosphere one degree.

WHAT IS THE GREENHOUSE EFFECT?

To answer this question, I first have to bring our favorite star into the discussion. Without the sun, there would be no point to greenhouses or metaphors about them. The sun is a great provider of energy, but sometimes it likes to mess with us. Not all of its rays are good for us.

We receive what I call the big three forms of radiation from the sun: the visible light that allows you to see what is going on, ultraviolet (UV) light, and infrared (IR) light. UV rays have the shortest and most energetic wavelength, while IR rays have the longest and weakest wavelength.

If you aren't already familiar with the greenhouse effect, it is a metaphor popularized by science presenters. Allow me to begin by describing how an actual greenhouse works. This peekaboo structure is designed to house plants and help them grow. The glass walls and roof allow sunlight to pass through, warming the air inside.

To paraphrase the wise words of a team of great thinkers known as Felder, Henley, and Frey (aka the Eagles), the light can check out any time it likes, but it can never *completely* leave. The UV rays are gobbled up by the grateful residents. The unused IR rays are reflected back up, but because of their puny wavelength, they have difficulty passing back through the glass and are now trapped inside and further heat the greenhouse.

Now substitute gasses in the atmosphere for the glass, and you can see how the effect of human activities warms the earth. A warm earth is a good thing. It is why life flourishes on this planet. Problems creep into the system when humans unintentionally play with the thermostat.

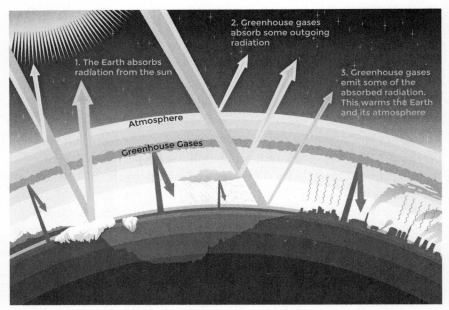

Fig. 11.1. Illustration of the greenhouse effect. (Modified from image by iStock Photo/DavidSzabo.)

PEOPLE WHO LIVE IN GLASS HOUSES SHOULDN'T THROW CARBON

One of the atmospheric gasses is carbon dioxide (CO_2), an all-star in the greenhouse heating game. It absorbs energy, mostly infrared energy, from the earth's surface and emits energy back down to the surface. Needless to say, the more CO_2 in the atmosphere, the more energy gets beamed back down to the earth's surface.

Both natural and human-made components create carbon cycles that shoot CO_2 into the atmosphere. By itself, the natural carbon cycle can lead to net cooling or heating. But in the last century, human activities have spiked the concentration levels to the point where now there is only warming. Humans release excess CO_2 when we burn fossil fuels such as coal or gas to warm a house, or power autos, or to generate electricity to play an Xbox.

Carbon is also released when trees are chopped down. A tree

absorbs about forty-eight pounds of CO_2 a year.[1] When it is cut, or rots, or is burned, all the CO_2 it has collected is released into the atmosphere.

Climate scientists are able to measure the weight of the carbon in the atmosphere over time. According to the National Oceanic and Atmospheric Administration (NOAA), the amount of CO_2 concentration in the atmosphere has increased from the 1880 pre-industrial level of 280 parts per million (ppm) to 400 ppm in 2016.[2] The last time the earth reached 400 ppm was during the Pliocene Epoch, between 1.8 million and 11,700 years ago, when the average temperature was two to three degrees Celsius warmer (almost four degrees Fahrenheit) than today.

Because of increased CO_2 concentrations, the earth's current average temperature has risen about one degree Celsius since 1880.[3] If no reduction in current emissions occurs, then the rise is on pace to increase six degrees Celsius by the end of this century.

HOW IS THE AVERAGE TEMPERATURE OF THE EARTH CALCULATED?

The French mathematician and physicist Joseph Fourier devised a way to calculate a planet's temperature. Because it was the 1820s, I will assume his planet of choice was Earth. He described it as the balance between energy received from the sun and how much was radiated back into space. In the 1850s, John Tyndall showed how atmospheric gasses help Earth to retain heat radiation.

According to the National Centers for Environmental Information (NCEI is a division of NOAA), as of January 2017, the average temperature of Earth was 12.88 degrees Celsius (55.18 degrees Fahrenheit). This is 0.88 degrees Celsius (1.58 degrees Fahrenheit) above the twentieth-century average for January.[4]

IS THERE EVIDENCE OF GLOBAL WARMING?

As always, look for evidence about any claim. In the case of global warming, the amount of evidence is vast. It includes rising sea levels, warmer oceans, a shrinking polar ice sheet, a declining artic sea, glacial retreat, and a decrease in snow cover. Going over all the evidence would be a book in of itself, so I'll elaborate on only a few.

CO_2 is a very light molecule. Imagine you take your significant other for a romantic date floating over Paris in a hot air balloon. Because the heated air in the balloon is warmer than the surrounding air, the balloon expands as the warmest air pushes to the top. The rising action lifts the balloon into the romantic night (and the possibility of you creating descendants).

This is similar to what happens to the lower atmosphere as it collects carbon. The lower atmosphere, called the troposphere, has shifted upward in recent decades. (Please note that this shift might have nothing to do with you having descendants.)

In 2009, the Bramble Cay melomys (*Melomys rubicola*; think big mouse) became the first confirmed mammal killed off by climate change.[5] They *used* to live on a low-lying island in Australia's Great Barrier Reef. The rising sea level destroyed 97 percent of the island's edible vegetation. No food, no melomys.

Since I brought up the Great Barrier Reef, let me tell you about reef loss. Today, about 75 percent of the world's reefs are at risk.[6] Coral reefs are a community of corals; coral is a polyp, little organisms with tentacles that snare plankton and other small creatures. There is truth in the saying, *it takes a village to make a reef.* Coral reefs are the feeding grounds for 114 species of fish and fifty-one species of invertebrates.

Menu prices are skyrocketing due to the warming and acidification of oceans. Warm waters cause coral to lose the symbiotic algae that produces its food, a process called coral bleaching. Researchers are working on restoring exiting reefs with tiny coral transplants. Other scientists are experimenting with breeding stress-tolerant corals.[7]

IS ANYTHING BEING DONE TO PREVENT GLOBAL WARMING?

Policies to mitigate global warming can range from taxing companies for polluting to some pretty drastic geoengineering proposals that include blocking sunlight.[8] Here is something to consider: who would control the policy on what the temperature should be? I don't know, but that would take some significant international cooperation.

The Paris Agreement, ratified in 2016, is an agreement to reduce carbon and to attempt to keep global warming from increasing another two degrees Celsius. This is good but not perfect. If the warming rises to this limit, plant and animal habitats will be destroyed, food crop yields will drop, and sea levels could rise.

If we want to get the job done (or at least slow it down), we need to do more than just lower emissions. We need to yank the existing carbon molecules out of the atmosphere. Enter carbon sequestration, the capturing and storage of CO_2.

A lot of methods can accomplish this, but like everything, it comes at a price. Most of the methods use huge amounts of energy, which could make the problem worse. Engineers could design artificial trees that suck CO_2 using chemicals that absorb and store carbon dioxide.[9] Another idea is to add algae to building facades to absorb CO_2.[10]

Don't forget oceans! We could fertilize them with large amounts of iron to bloom phytoplankton. Phytoplankton grows using sunlight and CO_2. We would have carbon sequestered under the sea. There is no shortage of ideas. Making them work economically is the problem.

Researchers at Cornell University have an exciting (and entirely possible) solution. They've designed an electro-chemical cell that captures CO_2 and generates rather than burns electricity.[11] Think of it as carbon sequestration battery. If they can get it from design to reality in a cost-effective and scalable manner, these batteries could be attached to carbon emitters like car tailpipes or factory smokestacks.

In Iceland, researchers have discovered a way to sequester atmospheric CO_2 by turning it into stone. They proved their concept by capturing emissions from a power plant and injecting it along with water into basalt rock. The gas formed into a very stable carbonate mineral with low risk of carbon leakage. A big hurdle of scaling this project is that it uses more water than the amount usually available to industrial sites.[12]

LET THERE BE (A LITTLE) LIGHT

One of the ways to prevent extinction from excessive climate change is to alter our physiology (transhumanism). This has already been covered in an earlier chapter, so let's jump to the alternative: modifying the planet. Geoengineering is all about changing the environment (rather than ourselves) to stave off the effects of global warming.

We could cool our planet by reducing the earth's absorption of sunlight. This can be done by emulating a volcanic eruption. It isn't as impossible as it might sound. If you have a pilot's license, you can join the fleet that conducts this operation. Squads of airplanes could fly into the upper atmosphere and spray sulfur. The permanent cloud would reflect sunlight.[13] The color of the sky would probably change, and weather patterns might go wonky, but less heat would reach the earth's surface and the planet would remain habitable.

Another geoengineering possibility is to create glacier cozies. Tarp these big guys with reflective covers to absorb heat and protect the ice below.[14]

Finally, how about a little sunblock? Space agencies could place trillions of satellite solar shields around the earth to reduce sunlight.[15] For this venture, we might have to mine asteroids to pull together all the necessary resources. We would also need some really good guidance systems to prevent them from crashing into each other.

WHAT OTHER HUMAN ACTIVITIES AFFECT WEATHER?

Besides burning fossil fuels, humans have inadvertently manipulated the weather in other ways. In general, human land use has changed evaporation rates and altered albedo, the proportion of light and heat reflected by a surface.

Our increase in urbanization has caused "urban heat islands," urban areas that are significantly warmer than surrounding rural areas. Cities tend to be warmer due to heat emissions from factories and cars. Buildings affect air flow, which redistributes precipitation in areas around cities. Metropolises cover land surface, lowering the albedo; also, when sunlight hits pavement, there is no evaporation and the sun's energy heats the ground.

Cities aren't the only places where humans affect the weather. Rural overgrazing and wind erosion results from poor agricultural planning. When trees (carbon absorbers and oxygen producers) are cut down to grow crops, surface albedo changes.

IS FICTION BEING USED TO HELP WARN ABOUT GLOBAL WARMING?

> *The Suck Zone. It's the point basically when the twister . . . sucks you up. That's not the technical term for it, obviously.*
>
> —Dustin (Dusty) Davis in the movie *Twister*

Yes, fiction helps spread the warning quicker, and to more people, than science journals. During the past decade, climate catastrophe has been used as a backdrop for science fiction and thriller novels. There is also eco-fiction, nature- or environment-oriented fiction like *The Perfect Storm* and *Twister*, that has its roots in science.

Sometimes fiction is threaded with a melodramatic urgency that belies solid science. Take the global warming disaster movie *The Day After Tomorrow*. In the movie, the Gulf Stream is shut down by global warming. This is not a real possibility, although mixed evidence does exist for its slowing. Also, the excessive rate at which the movie reports ocean rises from the melting of the Greenland ice sheet is exaggerated. The movie does succeed in spurring the general public to talk about the effects of global warming.

Tobias Buckell's novel *Arctic Rising* explores how the loss of ice in the Arctic Ocean might change international relations after the economic fortunes of nations are reversed.[16] One thing I'd bet on to *not* be fiction: if the environment keeps changing, we will end up with climate refugees. People would move to cooler areas that most likely are already occupied. In the novel *The Water Knife* by Paolo Bacigalupi, the water-depleted American Southwest is carved into small warring nation-states fighting over the remnants of the Colorado River.[17]

Let us not forget the aliens. Wesley Chu's *The Lives of Tao* is about aliens who want to settle on our planet. They deliberately harness humans to change the earth's climate by causing runaway greenhouse gasses.[18]

The Hugo Award–winning *The Three-Body Problem* by Cixin Liu artfully uses the backdrop of climate change to emphasize a political drama. Humans have failed the earth, so a sympathetic main character helps aliens take over because it is (possibly) what we deserve.[19] Michel Farber's *The Book of Strange New Things: A Novel* turns this idea on its head; humans fleeing a dying Earth cause human-style environmental havoc on an inhabited alien world.[20]

PARTING COMMENTS

Humans have inadvertently changed the atmosphere over the past century. The net result is global warming. This is an existential threat, but because it is happening so slowly, we don't feel any urgency to respond. This is a not-insignificant side effect of the survival instinct. The instinct, which has served us well, is to consume today rather than save for tomorrow. Therefore, many humans don't worry about the environment as much as they should.

Can we kick the fossil fuel habit? Swap it out for the low-emission cast of characters known as wind, solar, and nuclear power? Yes, we can, and we will. The only issue is the speed with which we act.

To give us the time we need to develop genetic and technological modifications (or habit modifications, or moving our rears ends out of town and to the stars), we can aggressively enact policies to reduce CO_2 concentrations. We might also use geoengineering technology

MAN AND NATURE 155

to slow the warming before the sea levels rise too much and gift us with floods, or before the oceans grow too acidic and tamper with our food supply. Let's just say that if you think parts of Africa are dry now, imagine them after droughts and extreme heat.

CHAPTER 11 BONUS MATERIALS

Bonus 1: Negative Greenhouse Effect

Rising CO_2 levels can sometimes lead to global cooling. Evidence of this is found in central Antarctica.[21] Rather than radiating heat from the ground, CO_2 increases the amount of heat that escapes into space. The amount expelled is usually offset by heat trapped in the ground. In Antarctica, the ground is so cold little heat is radiated.

This does not contradict the rise in the average global temperature. The increase in temperatures throughout the rest of the world overwhelms this small local increase.

Bonus 2: A Dawning of a New Age (Really, an Epoch)

The geologic timeline is divided into major divisions of geological time called eons. Eons are made up of eras, and eras are made up of periods, and periods are made up of epochs. We are currently in the Phanerozoic eon, Cenozoic era, Quaternary period, Holocene epoch.

Geological epochs are transitions that result in permanent change. The Holocene epoch began about twelve thousand years ago when the Ice Age receded and civilization arose. Good for us! This is our current epoch.

Or is it? A lot of scientists are calling for the announcement of a new epoch called the Anthropocene, the beginning of an exciting and scary time for us Earth dwellers. It is a period during which, for the first time in the planet's history, self-aware—meaning human-influenced—geologic forces can shape the planet. Steel, plastic, and concrete are this proposed epoch's techno-fossils.

Bonus 3: The Ozone Layer

Ozone is created from chemical reactions of the O_3 molecule in the upper atmosphere (stratosphere) as it graciously absorbs 97 percent to 99 percent of the sun's UV rays. Trust me, these are not the rays you want for your suntan. The DNA damage can lead to skin cancer. For every two million oxygen (O_2) molecules, only three O_3 molecules exist. When a UV ray hits an O_3 molecule, the ray splits it into an unstable O_2 that binds to another O_1.

If you want to mess with the chemistry of the ozone, send up some chlorofluorocarbons (CFCs). These release binding chlorine. A single chlorine atom can destroy up to 100,000 O_3 molecules.[22]

Not all of the environmental news is bad. The Montreal Protocol, which was adopted in 1987, was intended to phase out CFC use by 2010.[23] The initiative appears to have been a success. The ozone layer is on the mend after many years of bombardment with CFCs from refrigeration and spray-can propellants.

CHAPTER 12
TIME TO MOVE (PLAN B)

NASA are idiots. They want to send canned primates to Mars!
—Charles Stross, *Accelerando*

The universe offers many more ways to die than to survive. Surprise! Possibly the only place in the entire universe for us humans is right here on good old terra firma. Even on our home planet, we can't survive just anywhere. A lot of the land is frozen tundra or arid desert—and don't forget that most of the area is covered by oceans.

I wouldn't want to see the scary ending of the "all eggs in one basket" scenario. What if an asteroid struck the earth, or global climate changes caused human extinction (unless we embrace transhumanism), or, if you want to go all science fiction, alien invasion? So, if not solely for the pure joy of exploration, some of us should probably hightail it off the earth and start filling other baskets.

To me, the most interesting alternative to being cramped inside enclosed space station habitats or hollowed-out asteroids is to colonize another planet. Only, as with most things in survival science, this idea isn't quite that simple to realize. What are the odds of landing on some planet that offers the bare necessities and biodiversity essential for human life? Probably less than getting two royal flushes in a row.

I've heard that, on rare occasions, cheating arises in poker. An unscrupulous dealer might manipulate the cards to help one player. Since it is unlikely for humans to move to a ready-made planet, scientists have been thinking up ways of doing a bit of cheating to improve the odds. They have thought up ways to rig our bodies through genetic modification or to rig the planet's environment.

This chapter focuses more on the planetary science, but I'm sure genetic manipulation will pop in from time to time. Colonization will probably require a blend of both.

The climate science of the last chapter is important not only for understanding the earth but for understanding how to adjust the environments of other planets. As a science Zen master might say, "Before we can change another world, we must understand our own." This outward-looking planetary science is called *terraforming*, the science of making other planets (or asteroids) more hospitable to Earth life.

Terraforming is all about tweaking what a planet already has. The amount of tweaking depends on how different the biosphere of the target planet is from Earth's and/or on how much we have modified our bodies. In the end, terraforming might be our escape hatch if something disastrous happens to our planet.

Fig. 12.1. Illustration of terraforming.

Terraforming was first mentioned not in a scientific journal but . . . in science fiction. The word was first used in the short story "Collision Orbit" by Jack Williamson in 1942.[1] It took a while for mainstream science to get the jargon right. Does the original scientific term *ecosynthesis* sound as exciting? I thought not.

WHAT BASIC PLANETARY ELEMENTS MAKE FOR HAPPY COLONISTS?

1. They want what one of the Three Bears had
   The most ready-made planets for colonization are probably located in the circumstellar habitable zone of a solar system. Thanks to the popular press, you've probably heard this being called the Goldilocks zone, where a planet will be not too far from or too close to a star for our type of life. The greatest chance of a planet having sufficient atmospheric pressure to support liquid water on the surface lies within this zone.
   Outside the Goldilocks zone, water can exist, but it would probably have to exist below the surface of a planet or moon. It wouldn't be easily accessible. The zone's size varies between solar systems depending on a star's luminosity and age. And within a solar system, the zone doesn't remain constant. As our sun ages, it will bloat outward, pushing its own Goldilocks zone outward.
2. Flicking on the lights
   To make the planet conducive to life as we know it, specifically human type, there has to be enough light for photosynthesis. Plants need that energy to produce oxygen, and oxygen tends to be a good thing for colonists. To terraform a planet outside the Goldilocks zone, giant mirrors might be used to focus sunlight onto a planet. A higher-level civilization, maybe a type II, might consider dragging a planet closer to a sun. For now, this is pure science fiction.
3. The need for gravity
   The mass of the target planet is important. Remember that general relativity states that the more massive the planet, the greater the gravity. Colonists will be concerned about

gravity for at least two reasons. First, the planet needs enough gravity to maintain an atmosphere. Second, it directly affects the quality of human life.

The effects of microgravity (when an object appears to be weightless as in a freefall) include loss in bone density, brain damage, eye damage, and heart damage. Don't forget that your digestive system needs at least some gravity to help push down food and waste. Our bodies evolved in an environment of 9.8 meters/second^2 (32 feet/sec^2) of pressure. How low can gravity go before health is adversely affected? Scientists aren't sure.

On the flip side, scientists also aren't sure of the long-term effects of excess gravity. So, depending on the planet of choice, we must either alter the gravity to make it habitable or (again) alter our bodies to survive.

4. Take it out for a spin

The speed of planetary rotation matters. Too long or too short of a day can be challenging to baseline human life. We evolved on a planet with a twenty-four-hour day, so extreme day and night cycles might not be advantageous to our plant or animal life.

A couple of solutions are possible. We could use solar shades to manipulate a planet's day and night cycle. If you happen to come from a type I or II civilization, you might decide to change the rotational speed of a planet by crashing asteroids against it, thereby causing drag against its gravitational field to slow it down. Naturally, this is before settlement.

5. Magnetism

Our planet has generously constructed a magnetic field (called a magnetosphere) around itself to protect us from solar winds. Solar wind is a stream of charged particles blowing out from the sun at a rate of five hundred miles per second.[2] When solar winds strike the earth's magnetic field, the field stretches as it repels the winds and keeps our atmosphere intact. Finding a planet with one already constructed would be nice.

TIME TO MOVE (PLAN B) 161

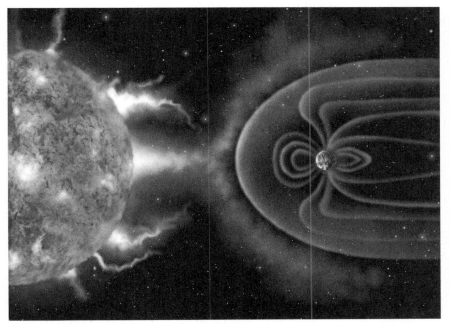

Fig. 12.2. Illustration of the magnetosphere. (iStock Photo/aaronrutten.)

Could one be added? In theory, yes. In terms of energy cost, it might not be worthwhile. A planet's core could be liquefied (like the earth's molten iron core) to create a planetary dynamo. The outer layer would rotate at a faster rate than its liquid center. How would you liquefy a planetary core? Many (many) nuclear bombs set to detonate around the core.

6. Tectonic plates

Colonists wouldn't want volcanic eruptions spouting out CO_2 at random times. Not to point any fingers, but if I did, my index finger would be aimed at Venus.

Tectonic plates are rigid plates in the planet's uppermost mantel. Subduction is the process where one plate is shoved beneath another. Subduction helps regulate Earth's natural CO_2 levels.

EXTRA RESOURCES: CAN ASTEROIDS BE USED TO OUR ADVANTAGE?

Quick fact: An asteroid is a large, rocky body in orbit around the sun. A meteoroid is a chunk of an asteroid, and a meteor is a meteoroid burning up in the atmosphere. A meteorite is what survives the passage through the atmosphere and whacks into the earth's surface.

It probably isn't a good idea to haul the earth's natural resources to the target planet. A better way to deliver raw materials is to mine asteroids. This idea first pops up in science fiction when we learn of Martians mining asteroids for gold in Garret P. Serviss's 1898 novel *Edison's Conquest of Mars*. I'm thinking that if exploiting asteroid resources was good for the fictional Martians, then it could be good (and possible) for us real earthlings. To keep the expense down, a multistep process called *optical-mining* would have to be used.

Step one is to design a vehicle that captures an asteroid in an inflatable bag. The spacecraft would have to be outfitted with reflectors that concentrate sunlight onto the asteroid. The heat would release water and other volatiles, chemical substances such as nitrogen or carbon dioxide. The best part is that the sun would be doing the mining, and for free!

Because mechanical drilling isn't required, maintenance would be cheap. Another benefit is that tons of water (a volatile) could be harvested and stored as ice. The water would be available for rocket propellant, used to produce oxygen, or even as drinking water for travelers.

Also, don't forget that hurling asteroids at a planet is not always a bad thing. Above, I mentioned that asteroid strikes can slow the rotational speed of a planet to make our day lengths more pleasant. As you will see below, asteroid impact can be used as a delivery system for volatiles.

THINKING GALACTICALLY BUT SEARCHING LOCALLY FOR A VACATION HOME

A snippet from *Futurama*:

> Fry: I'm impressed. In my time we had no idea Mars had a university.
> Professor Farnsworth: That's because then Mars was an uninhabitable wasteland, much like Utah. But unlike Utah, Mars was eventually made livable when the university was founded in 2636.
> Leela: They planted traditional college foliage. Ivy . . . trees . . . hemp . . . soon the whole planet was terraformed![3]

I suspect that this is not how Mars will be terraformed. The science sounds a bit shaky. I'm sure any colonist would be happy to locate a planet with the environment of Utah. Anyway, Venus and Mars make for some juicy local choices. As with most things, these choices come with their own unique set problems, but none are scientifically unsolvable. Although it might take a higher-level civilization.

The award for best size and location for habitation goes to Venus. It has 82 percent of Earth's mass and 95 percent of Earth's diameter, so gravity wouldn't be much of a problem at 91 percent of what you feel here. At its closest, Venus is only a half-shell away at a scant forty million kilometers (about twenty-five million miles).[4] Not too shabby, but I haven't told you the problems or, rather, the single big problem. The runaway greenhouse effect gasses make the atmosphere over ninety times thicker than Earth's, making it the hottest planet in our solar system. Did I mention it is also toxic?

All this means is that we have a fixer-upper. Ideas for terraforming range from seeding the atmosphere with hydrogen, to produce water that will rain down and create surface oceans, to mining magnesium and calcium on the planet for chemical carbon sequestration. As they have with the earth, geoengineers suggest putting up solar reflectors between Venus and the sun to cool the planet.[5] The panels would also help deflect the sun's solar winds, protecting the planet from radiation.

Venus has no magnetosphere, so it would be vulnerable to the solar winds without its thick atmosphere. As an alternative, it might be possible to place reflective balloons in the upper atmosphere. Why not have humans live in floating reflective cities while further terraforming is done?

If you asked the fictional Mark Watney from Andy Weir's *The Martian*, I'm sure he'd tell you that Mars is a solid possibility for settlers. Only make sure you don't go it alone. Mars is farther away than Venus, so the costs and risks involved of getting there are greater. At its closest orbital point, it is about fifty-six million kilometers away (about thirty-five million miles).[6]

Distance aside, Mars has a similar enough orbit compared to the earth's. Its day is only about forty minutes longer than ours. The planet contains both water and carbon. Mars is 53 percent the size of the earth, so visitors are stuck with only 38 percent of the earth's gravity. The lower gravity makes it hard to maintain an atmosphere. No magnetic field exists to protect colonists from solar radiation.

Back in the day (billions of years ago), Mars had a much more substantial atmosphere. Since then, most of it has leaked out into space. Thank the sun's solar winds. Because Mars is not protected by a magnetic field, solar winds blow off the planet's gas molecules at a rate of about one hundred grams of atmosphere every second.[7] When the sun is in a particularly foul mood, it bombards poor little Mars with solar flares, increasing the loss by a factor of ten.

Not having a magnetic field is a problem but not a scientifically insurmountable one. It might all come down to transhuman genetics or liquefying the planet's core. An alternate idea was proposed at the Planetary Science Vision 2050 Workshop hosted by NASA in February 2017. NASA discussed how positioning a magnetic shield at a specific location could create an artificial magnetosphere to wrap around Mars and protect the planet.[8]

Another headache is temperature. While Venus is too hot, Mars is too cold. Don't worry, an aspirin will help. Take (at least) two of the following ideas on warming and call me in the morning.

Carl Sagan thought it possible to transport dark materials that would reduce albedo to the planet's polar ice caps. There they would absorb more heat and melt the ice. NASA suggested introducing greenhouse gasses to create an oxygen- and ozone-rich atmosphere.[9] Volatile-rich asteroids could be aimed at the planet's surface where the volatiles would bolster atmospheric pressure. We could also try a reverse-Venus maneuver. There we had solar reflectors blocking sunlight; on Mars, we use orbital reflectors to direct more light onto the planet.

The Mars Trilogy by Kim Stanley Robinson does an amazing fictional presentation of the science of terraforming. The trilogy begins with the voyage to Mars and then follows the lives of generations of colonists. After creating and sustaining a habitat, they increase atmospheric pressure and temperature so that water can exist on the surface. This series has a lot going on from politics (within and between Earth and Mars) to the details to terraforming.

I would be remiss if I failed to mention other local options for colonies, namely the various moons of our solar system, including ours. Europa, Jupiter's largest moon, lies outside the Goldilocks zone but has water underground. To survive, colonists would have to live in permanently enclosed habitats or below ground.

With our own moon, we know that it is close, water sits beneath the poles, it's packed with helium-3 (an isotope we could exploit for safer nuclear energy), and we've been there before. At the very least, we could set up a base to mine the water and break it down into hydrogen and oxygen for use in rockets we launch to other destinations.

SEARCHING GALACTICALLY: WHAT IS AN EXOPLANET AND HOW DO WE FIND ONE?

Exoplanets are planets outside our solar system. Roughly 3,400 exoplanets have been discovered at the time of this writing.[10] They are pretty far away, but I like to think that someday they might be colonized during the expansion of the human empire. Only the government I envision wouldn't be as uptight as the Alliance in the Firefly film/TV franchise.

So, how do we find these planets? Because of the difficulty in detecting exoplanets with direct imaging (i.e., by seeing them), scientists use two highly successful indirect methods.

1. Transit method

    When a planet crosses in front of its parent star, the star's brightness dims a very small amount. The size of what passed in front of the star can be estimated from this decline in luminosity.

    In February 2017, the existence of seven Earth-sized planets about forty light-years out in the Aquarius constellation were announced.[11] Galactically, this is practically next door. All seven planets were found using the transit method. The closest three are just within the Goldilocks habitable zone. Their sun, Trappist-1, is only one-twelfth the mass of our sun and less than half as hot, so its Goldilocks zone is considerably closer.

    Over the span of only a couple of days, the innermost planet zips around the pygmy (Jupiter-sized) red dwarf Trappist-1. As a shout-out to Johannes Kepler, credited as the first person to determine that planets nearer to their suns orbit faster than planets farther away, I'll mention that the farthest planet orbits at a sloth's pace of twelve days.

    Astronomers measured the wavelengths of light blocked by each planet. Each gas has its own light wavelength, so eventually we might be able to determine which gasses are in the atmospheres of these planets. The exciting bit is that if oxygen is present, that might be the result of plant photosynthesis. Yes. I'm bold enough to say life, of the extraterrestrial kind.

2. Radial velocity (the wobble method)

    An exoplanet exerts a gravitational pull on its sun. Granted, the pull is miniscule relative to a star, but it does have a measurable effect. The force causes the star's orbit to wobble a wee bit away from the solar system's center. The bigger the planet, the greater the wobble.

    The Alpha Centauri system is composed from a triad of stars. The closest of the three is named Proxima Centauri. Circling it is a planet about 1.3 times as massive as the earth. This planet, named Proxima b, is only about 4.24 light-years away.[12] Evidence of this planet was discovered using the wobble method.

    The planet is pretty close to its sun, only 5 percent of the distance Earth sits from its sun. Its orbit is only 11.2 days. Despite the short year, it is potentially habitable.

I know—how is that possible? Well, Proxima Centauri is a red dwarf, a low-mass star. That makes it cooler than our sun, so the planet's nearness doesn't create a heat problem. The projected temperature of the planet is just about right for water to flow on its surface. In chapter 17 you will learn about a project to send unpiloted ships to Proxima b.

TO INVADE OR NOT TO INVADE?

> *Yet across the gulf of space, minds that are to our minds as ours are to those of the beasts that perish, intellects vast and cool and unsympathetic, regarded this earth with envious eyes, and slowly and surely drew their plans against us.*
>
> —H. G. Wells, *The War of the Worlds*

How would you like it if Martians invaded the earth and transformed our ecology to suit their habitation needs? I'm sure the humans in the 1890s had a lot to say, especially when Mars attacked—at least in H. G. Wells's *The War of the Worlds*.

I'll reverse the question. What if we find a planet with life already on it? Do we alter the planet to meet our needs? Would you terraform if you knew it might be environmentally disastrous to life that had evolved according to its own environment? The moral issue goes past intelligent or even sentient life; it reaches all the way down to microbial life that might someday evolve into intelligent life. The cost of terraforming won't be paid only by us. The cost will also be paid by the planet we homestead.

PARTING COMMENTS

> *It is always sad when someone leaves home, unless they are simply going around the corner and will return in a few minutes with ice-cream sandwiches.*
>
> —Lemony Snicket, *Horseradish*

Thanks to evolution, humans have a special relationship with the earth. Can we divorce it and pair up with a shinier and younger planet? Yes, but not easily. The science of changing a planet into a habitable version of the earth is called terraforming. Terraforming is adding to, or augmenting, what a planet already has. For some planets, it might be as simple as adding volatiles by dismantling a small nearby moon or as complicated as directing asteroids at it.

CHAPTER 12 BONUS MATERIALS

Bonus: How Do We Handle All the Radiation the Universe Throws at Us?

It might be possible to modify humans for space travel and inhabiting planets that don't have radiation shields. Our cells have the bad habit of undergoing decay and mutation when they are hit by radiation. And a lot of radiation fills space in the form of cosmic rays, high-energy particles that damage living cells. In fact, due to their exposure to cosmic rays outside the earth's magnetic field, the Apollo astronauts have been five times more likely to die of heart disease.[13]

A good start would be to find a way to protect our bodies from all that radiation. The solution might come from stealing a few genes from the microscopic invertebrates known as tardigrades. Because of their looks, they have been nicknamed the "water bear."

These creatures are capable of surviving extreme conditions including dehydration and the vacuum of space. Tardigrades also have a unique protein that shields its genes from radiation. When exposed, their genes do not split apart or mutate as would the genes of most other life-forms on Earth. The protein probably evolved as a defense during dehydration, which causes similar cellular damage.

And now for the cool part. Scientists have inserted the protein into cultured human kidney cells, boosting cell tolerance to X-ray radiation damage by 40 percent.[14] So it might be possible for the tardigrade protein to protect human DNA from the radiation of space. Someday our space explorers might have water bear DNA.

Fig. 12.3. Picture (illustration) of water bear tardigrade. (iStock Photo/Eraxion.)

CHAPTER 13
INTELLIGENCE COMES IN ORGANIC AND ARTIFICIAL FLAVORS

I'm sorry, Dave.

—HAL 9000 from *2001: A Space Odyssey*

Computers are wonderful devices that do so much for us. They are capable of storing, retrieving, and manipulating data or information. And unless I'm mistaken, you have one or three. I am so conditioned to using my technology that I get agitated at the mere thought of losing my smartphone. Like it or not, it doesn't matter. They are a part of our lives now.

We have reached the point where our computer algorithms are smart enough to mimic human problem-solving. It might not be long before they stumble upon flaws within their own coding and *choose* to self-improve. Think about it: programs looking into themselves for self-improvement. This sounds very human. It reflects intelligence, *artificial intelligence* (AI).

Before going artificial, let's start au naturel—as in biological intelligence. The human brain is wonderfully complex, much more so than any computer.

Cogito ergo sum (I think, therefore I am).

—René Descartes

But what am I?

—a possible human conscious thought

Humans might be the only species *currently* on the earth to wonder, what am I? Is your inner narrator the creation of neurons? Let's try to find out.

WHAT DO EMERGENT PROPERTIES HAVE TO DO WITH MY THOUGHTS?

An emergent property is when the sum of something's parts has a unique property that is not present in any of the individual parts. It triggers when there is enough of something to make something different. If you add more and more protons to an atom, you push it along the periodic chart and change which element it is. When you add more and more heat to water, you change its state. Speaking of water, it takes more than one H_2O molecule to make an object wet. Wetness is an emergent property that arises from combining more and more H_2O molecules.

These are all examples of cumulative processes. I have one more that you'll probably like: your brain. This organ is made up of billions of electrically excitable cells known as neurons. These neurons make trillions of connections by sending electrical signals along tendrils called axons. A few neurons can't make a mind, but if enough of them get together, voilà: consciousness.

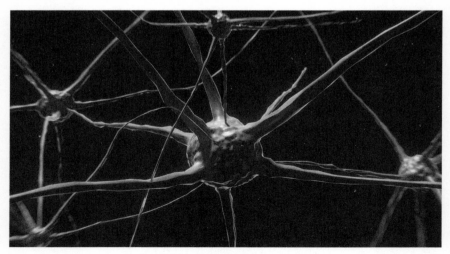

Fig. 13.1. Illustration of a neuron.

This particular emergent property (consciousness) is a side effect of evolution that allows the brain to sync with its environment. A lot goes on around us. If you come with me on a stroll through Central Park, your ears might register the sound waves from a dog barking while light waves project images to your eyes of a scrappy beagle. This input might activate memories of beagles you've known. All these processes occur in different regions of the brain, but they are synced to provide you with a unified experience of time and space. This is consciousness, or at least one of the definitions of consciousness.

What is really cool is that every brain on this planet has developed differently. Your brain cannot be repeated. You are so unique that the neural connections in your brain can be used as fingerprints for identification. A study out of Carnegie Mellon University found that even identical twins only share about 12 percent of the same neural patterns.[1]

Now, to blow your neurons by going all Jean-Paul Sartre (French existentialist philosopher) on you, human consciousness is a pretty good thing, but it carries an existential cost: knowing that all living things one day cease to live. We might be the only species that knows we will die. I'll let you in on something. I think that having human consciousness is worth the cost. Some people might disagree.

Summary: our brains are neurons flourishing and making connections. Useless neurons are culled by experiences; whatever isn't used is pruned. The gelatinous pack of neurons that remains somehow becomes your awareness. It can be introspective, like a feeling you get from smelling a flower, or how you react to different colors, or falling in love. It is our ability to have the sensory experience of a sore elbow after swinging a lightsaber.

HOW DO NEURONS CREATE CONSCIOUSNESS?

Different schools of thought debate the definition of consciousness.

One idea is that the brain constructs simulations of how the outside world works, and consciousness is this simulation. Two types of specialized cells found in a cortex of the medial temporal lobe give us a good sense of time and distance. One is a group of neurons called *grid cells*; they are the equivalent of a GPS system and are able to gauge the angle and speed of an object relative to a known starting point. These cells fire off regular signals as you move through space to form a mental map of the environment. The other group of neurons, called *speed cells*, allow the brain to update the map in real time.

Another school of thought is that consciousness happens when different parts of the brain connect and share information. This sounds a lot like when John Lennon defined life as what happens when you're in the middle of making other plans. Maybe he was a closet neurologist.

WHERE DOES COGNITION FIT?

Cognition is the higher-level function for processing comprehension. It is the drive to learn, to remember, to judge, and to problem-solve. Cognition interprets sensory input. If a tiger leaps at you, information travels from your eyes to your brain. Wisely (thank you, evolution), your brain signals your muscles to run.

To extract meaning from this incoming information, your brain must reduce excess sensory input. This is not a good time to notice the smell of jasmine or the gentle breeze on your face. Your entire attention is on the tiger, and your brain edits out unnecessary sensory information. It evolved to be reductionist.

Do you remember our park trip? I kept a log of where we went. After we left Central Park, we trekked through the city. Noise was everywhere, but you edited out most of it so you could concentrate on the conversation you were having with the street magician. You could hear other people talking, but your brain censored out what they said. Needless to say, my feelings are hurt. I was trying to tell you about cognition.

No matter how much you practice mindfulness (focusing on the present moment and the sensations around you), your brain will protect you from information overload.

WHAT IS INTELLIGENCE?

> *Intelligence is an accident of evolution, and not necessarily an advantage.*
>
> —Isaac Asimov

Intelligence is our cumulate information and skills. Nothing in nature dictates that human intelligence has been optimized. It is hampered by superstitions and short-term thinking (for example, our general inability to get hyped up about global warming because it is a long-term problem).

THE IMPORTANCE OF CREATIVITY (SOMETHING YOUR NEURONS MIGHT DO FOR FUN)

> *Creativity takes courage.*
>
> —Henri Matisse

Our brains are biological. They are very adaptable, which helps us survive in different environments and situations. Along the way, they developed creativity and a bit of lawlessness in society. Historically, rule breakers advance civilization. The universe was comfortably mechanistic after Isaac Newton laid down the foundations of classical mechanics. His laws of motion and gravity became gospel until that rule breaker Einstein used his imagination (via thought experiments) to work out special relativity.

The character Hari Seldon from Isaac Asimov's Foundation series took rules very seriously. He believed his fictional science of psychohistory allowed him to accurately predict the future . . . as long as everyone played by the rules. A successful project, until the rule breakers arrived.[2] It wouldn't have been much of a story if they had stayed home. Imagination is capable of shaking up complacent systems in real life just as in good science fiction. Take a moment to reread Arthur C. Clarke's first rule of prediction (found in the introduction).

Breakthroughs are unpredictable, but in the universe of computer algorithms, unpredictable does not exist. Computers are programmed to make predictions based on the prior behavior of particles, chemicals, or people. Breakthroughs defy forecasting. Comparing the brain to a computer is incorrect. It is not. So don't do it.

No mechanical or electronic law prevents an emerging AI from being more intelligent than humans. As I mentioned before, human intelligence can be inefficient. But will an AI ever be creative? Imagine an AI being a rule breaker.

RISE OF THE AI

Before there was AI, there was only human intelligence. In the early twentieth century, manual calculating, called computing, was considered "women's work." Mathematicians, possibly in an attempt to use their scythe-like wit to play on the word *man-hours*, described the computer output in "girl-hours." Henrietta Swan Leavitt was a computer in the early 1900's[3] when she discovered thousands of Cepheid variable stars that helped Edwin Hubble measure stellar distances (measuring distance this way was described in chapter 4).[4]

When mechanical computers came out during World War II, their calculating ability was measured in kilo-girls, a unit roughly equal to the calculating ability of a thousand women. The nonfiction book and subsequent movie *Hidden Figures* is about NASA's human computers. (By the way, the first person to write computer code was Ada Lovelace, daughter of poet Lord Byron. She did it on punch cards . . . in the 1840s!)

Since then, computing has developed to the edge of something more. Artificial intelligence is a computer system that can perform tasks that would normally require human intelligence. During our lifetime, machines have grown smarter because of human programming—because of human intelligence. Is there some critical mass of *smartness* where they can begin to program themselves? I wonder if this could be an emergent property.

Today's AIs can identify faces at an airport using photographs posted to social media, drive cars, or act as a translator like Arthur

Dent's babel fish from *Hitchhikers' Guide to the Galaxy*. Have you ever noticed that Facebook (if you have an account) knows your preferences and targets its advertising to you accordingly? It does. It has an AI scouring databases for your purchasing patterns. It also suggests other Facebook pages you might want to like. An AI can do all this and never grows tired or distracted. Or irritable.

All of the above are examples of weak AI. Strong AI is comparable to the adaptability of human intelligence. There aren't any pure examples of this yet. However, the program AlphaGo defeated the reigning European Go champion Fan Hui five games to one in 2016.[5] Unlike its predecessor Deep Blue, which relied on human coding to defeat chess champion Gary Kasparov in 1997,[6] AlphaGo learned to play from experience—meaning, as a typical human, it could be trained. It integrated two kinds of neural networks: the first to predict the next move and the second to evaluate the winner of each position.

SHOULD WE PUT IT TO THE TEST?

Alan Turing defined an intelligent program as one that could hold a conversation in a human language and convince a test subject that it was human. The test he devised, the so-called Turing Test, holds that if a computer can convince a sufficient number of judges a sufficient number of times that it is not a machine by answering a series of questions, then it is declared intelligent.

Imagine three rooms, each with a computer terminal. In room one sits a judge. She doesn't know who or what is in the other rooms, only that one has an AI candidate and the other hosts a human. After asking questions of her choosing, the judge must declare which room contains the AI. The problem with this test is, what counts as passing? If the judge gets it wrong more than 50 percent of the time?

The pure Turing Test is no longer used in modern AI research. Non-sentient computers are getting pretty smart, and humans can sometimes be tricked. A better test is needed. Perhaps the computer should discuss objects and people in context or perform tasks that require interpretation. Can the computer provide real-time commentary on the Super Bowl?

A test called the Lovelace Test, named in honor of Ada Lovelace, judges creativity.[7] Can the AI create an original work of art?

A machine doesn't have to be conscious to pass these types of tests. Consciousness is not a requirement for artificial intelligence. An algorithm does not need to be sentient to be effective. It does not need to experience subjectivity. An AI can effectively be a zombie and display just enough intelligence to act out its function.

That doesn't make AI consciousness scientifically impossible. A posthuman mind-space might have room for both downloaded human minds and conscious AI minds.

WHAT IS THE TECHNOLOGICAL SINGULARITY?

You cannot enslave a mind that knows itself.

—Wangari Maathai, Nobel laureate

To date, our technological progress has been limited by the intelligence of the human mind. There might (actually, there will probably) come a time when, with the help of computers, it will be possible to build a machine more intelligent than humans. This machine might be able to build an even more advanced machine, one that can rewrite its own coding software.

Self-programming for self-improvement is a recursive process wherein each generation of AI becomes more intelligent than the previous iteration. This machine evolution could occur over many versions beyond the original made by humans. Inevitably, there will come a time when humans will no longer be able to comprehend AI intelligence. It will be unpredictable. Perhaps, just perhaps, the first self-aware machine will be humanity's last invention.

The term *technological singularity* was coined by computer scientist and science fiction author Vernor Vinge. It marks the moment an AI triggers technological growth that is no longer comprehensible to humans.[8] This should not to be confused with the singularity of a black hole, but Vinge intended for that comparison to be made. During both types of singularity, there is a breakdown in our ability to predict what will happen beyond a certain point.

In the case of the technological singularity, the uncertainty is whether this intelligence will be helpful or harmful. Will we have created a god? Scared yet?

ETHICAL CONCERNS OF A POST-SINGULARITY WORLD

These questions are meant to get you thinking about the future, knowing that post-singularity humans won't be the ones answering them.

- Should an AI end global warming at the cost of jobs and a recession?
- What if ending global warming cost lives, say, a few thousand, but it *might* save millions of future human lives (they aren't yet born, so the births are hypothetical)?
- What if AIs could be bound to the rules of human law? If yes, which human laws? Should they be created by secular human institutions or follow human religious laws?

    Human organizations such as Google, Facebook, Microsoft, IBM, and Amazon are teaming up to develop an ethical framework for AI research. These overlords will protect us from the pitfalls of runaway AI technologies. They also keep a few on a leash to help us.

    Facebook is increasingly using AI to flag materials determined to be offensive by their policy and is working on a system to automatically detect fake news. Google is working on a set of tools to use machine learning to spot harassment and abuse. Its software package, called Conversation AI, will be able to detect hate speech.[9]
- Will there be a war over who gets to add these laws to the algorithm? Will an AI idly watch the conflict?
- What about love? Should it be (if possible) coded in?

    This idea has not been ignored by science fiction. In *A.I. Artificial Intelligence*, a movie based on a short story written by Brian Aldiss and brought to the screen by Stanley Kubrick and

Steven Spielberg, a Mecha (think android) named David is capable of projecting love and bonding with a single human. Is this cruel? How about when the human is gone and all that is left is a lonely immortal AI?
- After the (hypothesized) technological singularity, will many AIs exist or just one? If there are many, will they get along?

    The television series *Person of Interest* is about a battle between two AIs who use humans to wage their war over how the world should be run . . . and the destiny of humans.

FULL CIRCLE

I'll repeat something from the biological section:

> *Cogito ergo sum (I think, therefore I am).*
>
> —René Descartes

> *But what am I?*
>
> —a possible artificial conscious thought

Can a computer have its own inner narrator? Yes, but the *initial* narration must be provided by human intelligence. The notion of bicameralism (a two-chambered mind) originated in the controversial book *The Origin of Consciousness in the Breakdown of the Bicameral Mind* written in 1976 by psychologist Julian Jaynes.[10]

He believed that consciousness developed in humans only about three thousand years ago. Before then, humans ran more or less on automatic. This was the era of the bicameral mind, when the mind was divided into two parts. He proposed that when something novel was witnessed by primitive humans and simple habit or reflex wasn't enough to deal with the situation, a voice popped into their heads to provide commands.

This auditory hallucination was to be obeyed. The voice *might* have been believed to be an outside agent like a chief or a god. According to Jaynes, the shift from bicameralism to introspection was the beginning of consciousness, the awareness that no outside agent was putting thoughts into your head. It is when a person realizes the pronoun used by the voice is "I."

The bicameral mind hypothesis has not generally been accepted by mainstream science. Jaynes wrote the book about humans, but that doesn't mean it can't apply to burgeoning AI. In the TV series *Westworld*, inner narratives lead the protagonist android Dolores to consciousness.

PARTING COMMENTS

We evolved to stay alive, eat, and generate offspring. Somewhere along our timeline, humans developed a perceptual system to help with these tasks. After that, something strange happened. The brain began constructing a virtual reality rendition (cognition) of what it picked up through the senses. It considered what it saw or believed to be reality and gave it meaning.

Artificial intelligence is a computer system able to perform tasks that normally would require a biological (human) intelligence. Someday, a generation of computer systems might be able to rewrite their own coding software and then create a more advanced generation of computers. This is machine evolution, and it could lead to intelligences that don't need humans anymore. The point of runaway technological growth is called the technological singularity.

CHAPTER 13 BONUS MATERIALS

Bonus 1: Moore's Law

The human brain can store a petabyte of information. A petabyte is a million gigabytes or a thousand terabytes. Feel free to flash a smug smile at your computer's paltry terabyte hard drive. Sadly, there are biological limits to the size of the human brain. At this point, it might have (or might nearly have) evolved to the limits of its processing power. There are no such limits to computers.

In 1965 Gordon Moore, the founder of Intel, pointed out that the number of transistors on an integrated circuit doubles approximately every year (in 1975 he revised this forecast to a doubling every two years).[11] A transistor is an electronic on/off switch on a microchip. The more transistors a microchip has, the faster the processing speed. The faster the processing becomes, the more efficient the computer functions, at least as measured in terms of internal clock speed (meaning more calculations per clock tick).

Roughly speaking, a computer processor speed doubles every two years. This has become known as Moore's law. As a corollary, the price of computers halves. So far, Moore's law has proven to be pretty accurate.

Over the years, manufacturers have made transistors smaller and clocks faster so they can perform more computations per second. This method has a limit. Electronics get too hot if you force them to calculate too quickly. We therefore add more cores (groups of processors that calculate in parallel) to increase the computer's speed. Now if only we can perfect quantum computing . . . but that was a story from a different chapter.

Bonus 2: Time Is in the Eye of the Beholder

A correlation between pupil dilation and the perception of time appears to exist, at least for nonhuman primates. A study found that when pupils are large, monkeys felt time pass faster than it actually did.[12] The pupils might indicate how a person keeps track of time. Large pupils might reflect a chemically heighted sense of awareness from some shock, which triggers a need to mentally slow time.

When you are mellow, on the other hand, your pupils might be smaller, indicating that time is perceived as flowing faster. Of course this study was performed on monkeys, so the extension to humans is only conjecture.

Bonus 3: Nefarious Software (Three Annoying Computer Infections)

1. A computer virus is a code that replicates itself to corrupt a computer system. A computer virus must be run by a user to be effective.
2. Trojans do not replicate. Instead, they hide inside innocent-appearing programs just like the warriors smuggled away inside the Trojan horse of Greek mythology. If you run the Trojan program, it does its nefarious task, perhaps deleting files and opening other programs. Most likely, it is granting an intruder privileges to a system while appearing to be a harmless app.
3. A worm is a program that replicates on its own. Unlike a computer virus, it doesn't need a user to run it. It spreads across networks. A weaponized worm named Stuxnet was used to infiltrate and target Iranian nuclear plants.[13]

CHAPTER 14
THE RISE OF THE ROBOTS

Danger, danger!

—B-9 (Class M-3 General Utility Non-Theorizing
Environmental Control Robot,
aka, the robot from *Lost in Space*)

Joke: A robot walks into a bar, orders a drink, and lays down some cash. Bartender says, "Hey, we don't serve robots." And the robot says, "Oh, but someday you will."

I hope this chapter's opening joke isn't prophetic because that would really suck, at least for us biological types. Perhaps less so for robots. I don't know about you, but I'm not prepared to fight off a T-800 robot from *The Terminator*, so I suggest we defund the Skynet program. Now!

A robot can be defined as a mobile nonorganic device that is programmable and designed to perform functions traditionally done by humans. It is not a computer, although, like the T-800, an AI program can be housed inside a robot body. If an AI needs mobility, a robot is much more practical for downloading than a meat suit clone (a biological body).

Robots and AI are not the same thing. But they are the peanut butter and jelly of cybernetics as well as some frightening science fiction. With the exception of Teddy from *A.I. Artificial Intelligence*.

Robots and AI make up a *true* version of René Descartes's mind and body dualism. Descartes believed that the mind is not the same as the brain. The mind is the immaterial essence of being human. It is housed (for a time) in the body, but it is not a product of the body. As discussed in the last chapter, neuroscientists now believe our mind can be explained by the neuron connections of the brain. That means that the mind (intelligence) cannot be separated from the human body. An AI (mind), however, can be separated from the robot (body).

To be a practical helper, a robot needs autonomy, a way to move about and interact with the external environment. If we want a robot to perform detailed-oriented tasks like helping someone with a disability or performing medical surgery, it must be able to gauge how much pressure to use to open a door or lift a patient.

This type of robot cannot be created like an AI that masters a task after being programmed with millions of examples (as it can while learning chess or the game of Go). A robot not only must "think," but it has to interact physically with the world. To do that it must have an awareness of its surroundings.

Humans take their external cues from their senses. Robot sensors can be modeled on human ones. Cameras can be used for eyes, microphones for ears, a gyroscope for inner ear balance, and so on.

IN THE BEGINNING, THERE WAS A WORD AND A FEW WIRES

The word *robot* hit the science fiction universe in 1921 when Karel Capek introduced his play *R.U.R (Rossum's Universal Robots)*.[1] In this social commentary on the working class (*robot* is a Czech word for "worker"), artificial people are created to serve humans. The robots do not take too kindly to the idea of servitude and revolt.

Humanity had to wait another twenty years to be introduced to the word *robotics*. You can thank Isaac Asimov for this linguistic contribution. It appeared in his short story "Liar" published in *Astounding Science Fiction*. Asimov claims he didn't know he originated the word. He thought it already existed as the equivalent of mechanics for machinery.[2]

If not in word, the idea of robot mechanized workers can be traced back as far as ancient Greek mythology. Talos, an automaton made of bronze forged by Hephaestus, protected Crete from pirates. Less poetic, although a lot more scientific, is the Greek Antikythera mechanism discovered in 1900. The device dates back to between 205 BCE and 100 BCE. It is believed to have been a computer of sorts that predicted astronomical positions and eclipses.

Around 60 CE, Heron of Alexandria designed mechanical devices for lifting and transporting heavy objects. Not all the automatons created by the Greeks were necessarily practical. Some were made for intellectual pursuits. The Greek mathematician Archytas built a steam-powered robo-pigeon sometime between 400 and 350 BCE to research how birds fly.[3]

In 1206 CE, Ibn al-Razzaz al-Jazari, the chief engineer at the Artuklu Palace in Turkey, completed *The Book of Knowledge of Ingenious Mechanical Devices* in which he described hundreds of mechanical devices.[4] Not everything he made was intended to be useful. One of his creations was a musical mechanical band.

At the 1939 world's fair, Westinghouse debuted Elektro, a seven-foot-tall mechanical man that could walk, talk, and perform a final activity that was appropriate for the time—smoke a cigarette.[5]

Personally, I find Elektro a bit frightening. I think that is why in 1940 he was joined by a dog-shaped companion named Sparko.[6] I checked and found no evidence of Elektro ever having revolted against his human creators.

Ford Motors accomplished two famous firsts in robotics. The first industrial robot was used by them in 1961. And according to the *Guinness Book of World Records*, the first human death by robot took place at the Ford Motor plant in Flat Rock Michigan on January 25, 1979. A mechanical arm struck Robert Williams a fatal blow to the head.[7]

ROBOTIC EVOLUTION

The evolution of robots isn't so much about its *form* as it is about its *status*. Initially robots were created as automaton slaves under the direct remote control of a human. This type of robot is increasingly popular in soldiering. Highly trained pilots fly drones all over the world. As of 2014, the United States military had over ten thousand recorded drones in its service.[8] Robot ground vehicles are used for defusing improvised explosive devices (IEDs).

188   BLOCKBUSTER SCIENCE

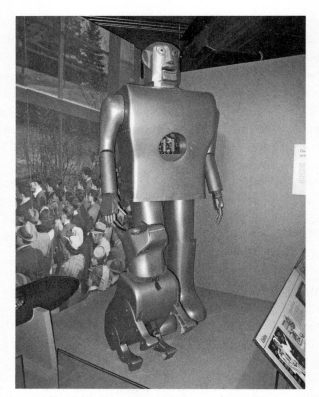

Fig. 14.1. Illustration of Elektro and Sparko.
(Wikimedia Creative Commons, author: Daderot.)

All of these rely on outside agency to perform their duties. This means they rely on an outside manipulator to control their actions (a slave to another's will). If we add some basic programming to allow for a semblance of autonomy, robots go from slavery to servitude. These can be self-driving cars or the Roomba that cleans up after you. Finally (possibly), they become coworkers with sensory input and weak AI. After that, I leave it to science fiction to speculate.

TO SERVE AND OBEY

Asimov and his robot fiction hold a special place in the science fiction cannon. He understood the potential danger of robots, and, as reported in the short story "Runaround" for the March 1942 issue of *Astounding Science Fiction*, he introduced the handbook of robotics, fifty-sixth edition 2058 CE. To save our collective human butts (at least in his science fiction universe), he introduced the now legendary three laws of robotics, which made them subservient to humans. He also thought up the pliable positronic robot brains that are force-fed these laws.

Isaac Asimov's classic Three Laws of Robotics (1942):[9]

> Law One: A robot might not injure a human being or, through inaction, allow a human being to come to harm.
> Law Two: A robot must obey the orders given by human beings except where such orders would conflict with the First Law.
> Law Three: A robot must protect its own existence as long as such protection does not conflict with the First or Second Laws.
> Asimov added a zeroth law in the 1986 book *Foundation and Earth*.[10] This law is of the highest order and supersedes the First Law.
> Law Zero: A robot might not injure humanity, or by inaction, allow humanity to come to harm.

Things get tricky if we were to actually force these laws on robots. In many of his stories, Asimov demonstrated how the rules lead to dilemmas. In his later books, adding the zeroth law actually forced robots to stunt human development by preventing the species from taking any collective risks.

As robots become more intelligent (adding in weak AI), communication with humans might lead to dilemmas. Human communication is nuanced. To the uninitiated, the intent of our words might be vague. I'm talking about a robot trying to separate how we ask for something from what we really want.

The misuse of a metaphor could be disastrous. Imagine after a tough day at work you causally say, "My boss is killing me." What should your robot think about this? Does it contact the police? Should the robot murder your boss in order to protect you?

Roger Clarke, a consultant in information technology, addressed this problem when he updated Asimov's laws. He also added sections on how robots should behave in large numbers and how to behave in robot hierarchies.[11] Here is Clarke's Extended Set of the Laws of Robotics:

> The Meta-Law: A robot might not act unless its actions are subject to the Laws of Robotics.
> Law Zero: A robot might not injure humanity, or, through inaction, allow humanity to come to harm.
> Law One: A robot might not injure a human being, or, through inaction, allow a human being to come to harm, unless this would violate a higher-order Law
> Law Two:
> (a) A robot must obey orders given it by human beings, except where such orders would conflict with a higher-order Law.
> (b) A robot must obey orders given it by superordinate robots, except where such orders would conflict with a higher-order Law.
> Law Three:
> (a) A robot must protect the existence of a superordinate robot as long as such protection does not conflict with a higher-order Law.
> (b) A robot must protect its own existence as long as such protection does not conflict with a higher-order Law.
> Law Four: A robot must perform the duties for which it has been programmed, except where that would conflict with a higher-order Law.
> The Procreation Law: A robot might not take any part in the design or manufacture of a robot unless the new robot's actions are subject to the Laws of Robotics.

Gordon Briggs and Matthias Scheutz of Tufts University set up a test using software designed to enhance a robot's natural language capabilities by improving the robot's conceptual framework.[12] A set of questions helped the robot decide whether to carry out a human command. Now a robot can simulate a pout and stomp its foot, declaring *no* to any command that might lead to contradiction. A bit of disrespect can avoid a lot of difficulties.

THE RISE OF THE ROBOTS   191

Question 1: Do I know how to do X?
Question 2: Am I physically able to do X?
Question 3: Am I able to do X right now?
Question 4: Am I obligated to do X based on my social role or relationship to the person giving the command?
Question 5: Does it violate any normative or ethical principle for me to do X, including the possibility I might be subjected to inadvertent or needless damage?

Unlike Asimov's and Clarke's laws, which are designed to protect humans, robotics physicists Mark W. Tilden developed a set of programming principles intended to nudge robots toward sentience. He believes that humans can take care of themselves. These principles are not much different than the evolutionary imperatives of being human. In fact, if you substitute the word *human* for *robot* in the principles, you get a hierarchy for humans. His three guiding principles for robotics:[13]

Principle 1: A robot must protect its existence at all costs.
Principle 2: A robot must obtain and maintain access to its own power source.
Principle 3: A robot must continually search for better power sources.

The condensed version:

Principle 1: Protect thine ass.
Principle 2: Feed thine ass.
Principle 3: Look for better real estate.

ROBOTS IN OUR EVERYDAY LIVES (AS SERVANTS)

1. Robots to do your driving
    To avoid the thousands of crashes annually, the US Transportation Department has proposed that all new cars be able

to communicate with each other about their relative locations and speeds using wireless technology.[14] Of course, the car manufactures would have to ensure that their vehicles speak the same language as the vehicles made by their competitors. Good luck.

Robots are all about trust. It takes a lot of trust to be a passenger in an autonomous car. This is where Uber pops in. The company is rolling out these autos for its customers. A fleet of Uber autonomous cars has been sent to Pittsburg and cities in California. These cars limit their speed to twenty-five to thirty mph, unless it is safer to match the speed of traffic.[15] Google has been trying for some time to get humans out of the driver seat; the company has been working on its own version of self-driving cars.[16]

2. Robots used in medicine

The development of movement in children with brain damage or cerebral palsy is often delayed. Their brains won't build and reinforce the connections involved with motor skills. Enter robots.

Researchers at the University of Oklahoma have developed technology that promotes crawling.[17] They came up with a tech-onesie and a robot on wheels loaded with a machine learning algorithm. The robot supports the baby, detecting kicks or weight shifts, and rolls in the indicated direction. This gives the baby practice crawling while stimulating the brain's motor control areas.

Robotic tele-surgery might allow surgeons to remotely manipulate robotic arms. Specialists from anywhere on the planet could perform life-saving procedures anywhere else. They could watch their actions using a 3-D display of the operating theater. With the proper feedback enhancements, surgeons could feel what the robot arms touch in real time.

Have you heard about Embodied Cognition in a Compliantly Engineered Robot (ECCEROBOT)? This robot was built at the Technical University of Munich to allow researchers to study how the brain and our movement interact.[18] Built with human-made bones, muscles, and tendons (human-made as in artificial; not human-made as in from a human body), ECCEROBOT reveals how the human neural and anatomical system moves.

Are you stressed out or concerned about a sick family member? Help is available in four feet of cuteness known as Pepper. This emotional robot uses facial and voice recognition to determine a person's emotional state ranging from sadness to concern. Aldebaran Robotics and SoftBank introduced the humanoid-shaped robot in Japan in 2014.[19]

I can't imagine people treating this little guy (thing?) like an appliance. As long as humans are capable of forming emotional attachments, we can't help but connect to robots, at least the cute ones. Pepper learns a user's personality traits and adapts to the person's habits, placing us one step closer to *Westworld*.

Finally, the CardioARM is a robotic surgical system designed for cardiac surgery.[20] With its articulate design, it snakes through the chest and wraps around the heart to perform delicate surgery without the need to crack open the patient's chest. A doctor controls it with a joystick.

3. Robocops

A robocop is no longer science fiction. The Robocop in the movie of the same name was actually a transhuman cyborg, but the Knightscope (K5) is a true robot.[21] The five-foot, three-hundred-pound K5 prototype is equipped with all the expected cool equipment like a 360-degree video camera, thermal imaging, laser range finder, radar, and a microphone to pick up sound. All that technology notifies authorities of crimes being committed in schools and businesses. K5 can be found patrolling the Stanford Shopping Center in California. So if you go there to do a little shopping, behave.

A question to ponder: should robots be equipped with a weapon, even a nonlethal one? I wonder how easy it is to distinguish between a twitchy customer and a gunman. If you are uncomfortable with an armed robot patrolling a mall, then how would you feel about using one in the military?

QUESTION OF ETHICS

It is time to wonder what happens when the peanut butter and jelly get together. Until a strong AI is developed, and most probably after, intelligent machines will remain emotionless (unless there are adrenaline algorithms) and mathematical. Put this AI into a robot, say, a war-drone, and send it out to hunt enemies. The drone will use cold calculations to search and destroy an adversary based on a pre-programmed probability. What should that threshold be? Fifty-one percent? Eighty percent?

As robots become more competent at decision making, what is the moral limit to what we can demand of them? If you program a robot to act happy, how different is this from when you cheer up a friend? Is a simulated emotional response very different from a "real" human reaction?

PARTING COMMENTS

To begin with, I'm happy to say that, as of May 2017, two operational robotic rovers named Opportunity and Curiosity are puttering along the surface of Mars doing science stuff. Robots do so much for us. Humans started out controlling machines remotely, but as we made incremental changes and conceded more control to the machine, we ended up with autonomous cars driving us to malls where we can be protected by K5 the robocop. The future sneaks up on us.

The evolution of robots in human society will be slavery, servant, coworker, and master. Okay, the last stage was a joke. I'm sure our benevolent friends will help humanity.

In science fiction, the stories of Isaac Asimov in particular, misunderstandings and consequences can arise from blindly accepting human commands. Do you think this could happen in real life?

CHAPTER 14 BONUS MATERIALS

Bonus: A Few Celebrity Science Fiction Robots

Gort (*The Day the Earth Stood Still*)
R2-D2, C-3PO, BB-8, K-2SO (Most things *Star Wars*)
Robby the Robot (*Forbidden Planet*)
T-800 (*The Terminator*)
Futura (*Metropolis*)
Marvin (*The Hitchhiker's Guide to the Galaxy*; apparently, robots can be depressed)
Delores and Wyatt (*Westworld*, the TV series)
TARS (*Interstellar*)
Number 6 (*Battlestar Galactica*)
The Robot (formally known as B-9, *Lost in Space*)
Johnny 5 (*Short Circuit*)
David and Teddy (*A.I. Artificial Intelligence*; I have a soft spot for helper teddy bears)
Tobor (*Tobor the Great*; the creativity of spelling robot backward, alone, gets Tobor on my list)
Mechagodzilla (various *Godzilla* movies; science may not be able to explain Godzilla, but there is hope for this guy)
The Gunslinger (*Westworld*, the original movie; I like Yul Brynner)
Twiki (*Buck Rodgers in the 25h Century*)
Bender (*Futurama*; included because of his positive outlook)

CHAPTER 15
ARE WE ALONE? EXTRATERRESTRIAL INTELLIGENCE

Klaatu barada nikto.

—Gort, robot from *The Day the Earth Stood Still* (1951)

In the 1940 short story "Farewell to the Master" by Harry Bates, we are introduced to Klaatu and his companion robot Gort.[1] This first contact wasn't pleasant, especially for Klaatu, who is shot. If you haven't read the story, you might have seen the movie based on it, *The Day the Earth Stood Still* (original 1951, remake 2008).

Both ask an important question: could the world's governments create a unified policy if extraterrestrials popped in for a visit? Or might some governments decide to "shoot first and ask questions later"? Would this shot be interpreted by the aliens as representative of the world?

The original movie was made during the Cold War when politics dominated and science fiction aliens represented our worst fears of outsiders. *The Day the Earth Stood Still* played against this theme by calling for an end to global hostilities. If you have not read the original short story, I suggest you give it a try. Its twisted surprise ending is very different than the movie treatment.

Many other examples of first contact with aliens occur in science fiction. In the movie adaptation of Arthur C. Clarke's *2001: A Space Odyssey*, alien black monoliths make first contact with hominins. A dialog between civilizations isn't as important as supercharging hominin evolution; eventually future humans are guided to head toward Jupiter. In the 1997 movie *Contact*, communication is established using prime numbers to translate messages. In *Close Encounters of the Third Kind*, a tune was used in the belief that melodic harmonics don't change across the galaxy.

Fig. 15.1. Illustration of *The Day the Earth Stood Still*. (Twentieth Century Fox.)

The 2016 movie *Arrival* plays on the theory of linguistic relativism known as the Sapir-Whorf hypothesis. Linguistic relativism proposes that language shapes our thoughts, or at least influences them. When we learn a new language, therefore, our thinking can be rewired. The lead character learns bits of the alien language by acting out words and decrypting written replies. The more she learns, the more her thinking changes.

Who says communication with alien cultures has to be written or verbal? What prevents aliens from attempting communication through scent or touch?

Sometimes first contact is last contact.

—Captain Janeway, from the *Star Trek: Voyager* episode "Waking Moments" (first broadcast January 14, 1998)

Whatever the method of communication, the world needs a protocol for first contact with an extraterrestrial expedition. There is one, sort of. Read the International Academy of Astronautics' proposed "Declaration of Principles Concerning Activities Following the Detection of Extraterrestrial Intelligence."[2] It is included in a chapter bonus.

NEVER MIND MEETING THEM. DO ALIENS EXIST?

Are we alone? I'm talking big picture here, not just you and me. I'm asking if the earth is the only place in the universe with intelligent life. Of course, more than one definition of *intelligent* life is possible. I'll stick to the definition of intelligence we used in chapter 14.

Here is a second question to ponder: if intelligent life does exist out there, will we be able to communicate with it? In science fiction, it's fun to shout, "Yes!"

A lot of science is involved in the potential answer, along with a lot of math. I know, I promised you early in this book that there would be no math. Don't worry, you won't need any scratch paper. I will do most of the work.

Of the two views on the possibility of intelligent communicating life outside our solar system, both rely heavily on statistical probability. The first group shouts, "Definitely, yes!" The second group taps the first's shoulder and says, "Don't be so sure." First I will work out the numbers for the group telling you, "Heck, yeah." Then I'll give you the other side of the argument. Both arguments are scientifically consistent.

The affirmative view comes from the famous Drake equation, which calculates the number of alien civilizations that can communicate with us right now. Professor Frank Drake wrote out the equation at the 1961 SETI (Search for Extraterrestrial Intelligence) meeting. You can look up a lot of fancy notations related to this, but I'll break it down for you in list form.

1. Begin by calculating the number of stars in the universe. The observable universe is estimated to have about one hundred *sextillion* (one hundred followed by twenty-one zeros) stars. It's probably best to focus on the Milky Way, which has a paltry two hundred billion stars (a low estimate).[3] Because of the vast distances between galaxies, communication between them requires some very interesting science fiction to explain. The distance between the Milky Way and the Andromeda galaxy is estimated at three million light-years.

2. Calculate the fraction of stars in our galaxy that have planets in their orbits. Let's be conservative and say 20 percent have planetary systems (20 percent of two hundred billion equals four hundred million stars with planetary systems).
3. Calculate the fraction of planets that can sustain life. Let's go with one planet for each star calculated in #2 above.
4. Calculate the fraction of life-sustaining planets where life has evolved. (Note that where life does occur, 100 percent of it has evolved.) Let's estimate that at 50 percent.
5. Calculate the fraction of planets where life has evolved intelligence. I'll assume 30 percent of planets where life exists have evolved intelligent life.
6. Calculate the fraction of planets where intelligent life has the ability for interstellar communication. I'll posit 10 percent to be conservative.
7. Calculate that civilization's longevity. What fraction of communicating civilizations are still around to set up a Twitter account? For example, they might have been able to communicate with us ten thousand years ago, but we weren't listening back then. Are they still around today? Do you think our civilization will still be able to send messages into space ten thousand years from now? As posthumans, would we want to? Let's set the chance of two civilizations existing at the same time, and being willing to communicate, to one in ten thousand.

Using these estimates, the Drake equation calculates six hundred communicating extrasolar civilizations in our galaxy. Feel free to tweak (or even tweet) some of my numbers and see what you come up with. My conclusion is on the low end of estimates.

SO, WHY HAVEN'T WE HEARD ANYTHING FROM ALIENS?

This question is known as the Fermi Paradox. Enrico Fermi was a physicist who received the 1938 Nobel "for his demonstrations of the existence of new radioactive elements produced by neutron irradiation, and for his related discovery of nuclear reactions brought about by slow neutrons."[4] His paradox addresses the contradiction between the high probability of extraterrestrial life capable of communication (as determined by the Drake equation) and the absence of evidence for such life. This is pretty important. After all, the universe is very large and pretty old by our standards, so why haven't any advanced civilizations tried to contact us?

The science fiction universe wouldn't be much fun if authors accepted the simple answer: there is no contact with intelligent extraterrestrial life because it doesn't exist. So you can see why we need scientific reasons for the lack of contact (as well as proof that they exist).

It's possible they don't know about us because we are too far away. Our television and radio signals have only been leaking into the cosmos for a century. Traveling at the speed of light, the farthest they have spread is about one hundred light-years from home. The Milky Way is 100,000 light-years wide. It's more than possible that our radio signals haven't reached them yet. It could take thousands of years before they can binge watch the human-zombie series *The Walking Dead*.

Another possible answer to Fermi's paradox is the theory that another livable galaxy might have developed only recently. As it turns out, harsh environments are not conducive to life. Surprised? Back in the (relatively) old days, the universe was a lot smaller and more chaotic. Gamma-ray bursts from the hot and energetic activity might have sterilized any newly formed planets.

Only after the universe had expanded enough and galaxies were far enough apart did regions of space that were safe for life emerge. Planets like Earth that reside near the galactic edge would have breathing room for life. If this is true, few extraterrestrial civilizations would have flourished before Earth's. It is possible we are one of the first. The newbies who are too young to send signals might be wondering if life exists off their own planets.

PERHAPS EARTHLIKE PLANETS ARE RARE

The anti-Drake(ish) argument, called the Rare Earth Hypothesis, seeks to demonstrate how unlikely the existence of earthlike planets—i.e., populated by complex intelligent communicating life—might really be. As before, I will present the argument in list form. And it all begins with a star.

1. An earthlike planet needs a host star in the galactic habitable zone. If the star is too close to the galactic center, radiation and black holes will cause problems. If it's too far away from the galactic center, too few metals will be available to form rocky planets. Approximately 10 percent of the stars in our galaxy appear to reside in a galactic habitable zone.[5]
2. The host star must be a singleton. Life under a binary system is tricky. I'm not saying impossible, but it is very unlikely. That being said, a long, long time ago, the population of Tatooine from *Star Wars* thrived beneath two suns.
3. The host star has to exist long enough for life to form on the orbiting planets. Our solar system has a nice 4.5-billion-year-old G-type star. Only about 7 percent of the stars in the Milky Way are G-type under astronomy's stellar classification known as the *main sequence*.[6] An ordered series of letters (O, B, A, F, G, K, and M) are based on brightness and color. O represents the hottest type of star, while M sits on the other end wearing shades (it is totally the coolest of stars). In case you are ever tested on this, remember this mnemonic device: Oh Be a Fine Girl/Guy, Kiss Me.

    About one billion years passed before life kicked off on our planet. Our sun should make it to the ripe age of ten billion. Don't pop the cork yet because we don't get to use that entire time period. We must leave the planet within a billion years or risk extinction. But that is a topic of another chapter.

    Now that we know what kind of sun is needed for life, let's consider life's planetary needs.

4. As we currently understand life, liquid water must exist for all the necessary chemical reactions. Therefore, an earthlike planet must have water. See bonus 2 of this chapter for details.
5. The earthlike planet must have a magnetic field to protect life from cosmic rays. Mars doesn't have such a field, which might be why we aren't regularly invaded by Martians.
6. The earthlike planet must have plate tectonics to circulate carbon. A chapter bonus will explain the importance of carbon and the carbon cycle.
7. Another large planet needs to hang around and deflect meteors attracted by the sun's gravity. This planetary guard must be nearby, yet it must be far enough away to avoid affecting the earthlike planet's orbit. For us, Jupiter plays this supporting role. It protects us from asteroids. Most of the time, anyway (sorry, dinosaurs).
8. For intelligent life to develop, the planet probably requires a long period of stable climate formed by a stable orbit and rotation. So, the earthlike planet needs a large moon to hold its rotational axis constant. Without that, a planet's axis can vary widely.

    For a moon as large as ours (relative to our planet's size) to exist is quite rare. Our moon (science believes) was created after a Mars-sized planet collided with Earth. The assaulting planet is named Theia for the Greek goddess who mothered Selene, the goddess of the moon.

Considering all the Rare Earth conditions, I estimate a one-in-a-billion chance for an earthlike planet to exist. So yes, Earth is indeed rare. Now for the important bit: my estimate is only for the *existence* of habitable planets. I have not calculated the chances that intelligent life actually evolved on any. I'll leave that to you. I suspect that only a fraction of the one-in-a-billion rare earthlike planets will have such life, perhaps only one or two per galaxy.

ANY IDEAS ABOUT WHAT ALIEN LIFE LOOKS LIKE?

Does *how* an alien looks matter to you? Would you be more trusting of aliens if they looked similar to us? Or had wide, sad puppy eyes? Sometimes the extraterrestrial life in science fiction is implausible because not only are they bipedal hominins, but they are somehow able to mate with us. I'm not arguing against evolution favoring bipedalism in intelligent life, but science fiction, television in particular, needs more biotype diversity!

So, what could an alien really look like? Probably a single cell, but if you are talking about intelligent life, that depends on planetary conditions and evolution. It's very possible they might have two eyes for binocular depth vision, which might be a standard survival trait everywhere. At the very least, they would evolve a light-sensitive, eyelike organ.

They will probably have a metabolic system to generate energy from some type of fuel, and a waste system to dispose of what isn't needed. In most cases, life on Earth appears biased toward symmetrical creatures: two arms, eight arms, etc. Only in rare exceptions do Earth species have an odd number of limbs (e.g., the starfish).

One could possibly be the loneliest number (at least according to the Three Dog Night song), but single-limb species like snails or worms are still considered symmetrical. Will this symmetry hold for species on distant worlds, or will three legs and three arms offer better evolutionary advantages in an alien environment?

Another idea not excluded by science is that life on other planets might be dominated by an element other than carbon. Aliens could be what we would consider rocky monsters composed of silicon atoms. In a *Star Trek* episode, thanks to the wounded silicon Horta from Janus VI, we learned something very important about the doctor of the USS *Enterprise*. To quote the good Dr. McCoy, "I'm a doctor, not a bricklayer."[7]

Silicon atoms have the same number of electrons in the outermost shell (*valence* electrons, which are used to form chemical bonds), so they are able to form bonds similar to carbon, except they aren't as stable. Silicon life would experience respiration and waste concerns. Carbon-based life excretes CO_2 gas while the equivalent for silicon is SiO_2, and it isn't as easy to unload. It's a solid. Talk about exhaling chunks.

Something to ponder: The majority of the matter in the universe is dark matter, something we can't see or interact with directly. What if an alien race could interact with dark matter as easily as we do ordinary matter? Would we even be able to see them? How would this affect their evolution?

PARTING COMMENTS

> *There might be aliens in our Milky Way galaxy, and there are billions of other galaxies. The probability is almost certain that there is life somewhere in space.*
>
> —Buzz Aldrin

This chapter has been about the chances of communicating with intelligent alien life. If the chances are truly high, why haven't they made first contact? I want to leave you with one more reason we might not have received their voicemails. If a large number of intergalactic civilizations (or even only a few) exist, we might be out of sync with them. Here's why.

Detecting an extraterrestrial intelligence's signal during the fleeting time in which the sender still exists is unlikely. Intelligent life could have developed and burned out long before we decided it would be fun to export television signals of *Twilight Zone* episodes. After we've transitioned into our post-biological form, it's very likely that we will no longer consider it worthwhile to send radio signals because we'll be operating in an expanded virtual universe (chapter 21 covers virtual reality).

CHAPTER 15 BONUS MATERIALS

Bonus 1: Ufology

Ufology is the study of reports, records, and physical evidence related to unidentified flying objects (UFOs). Don't let the -olgy fool you. Ufology is pseudoscience, a belief mistakenly taken as science. Nothing about it is based on the scientific method. Hopefully, this book will help you separate the real from the pseudo.

Pseudoscience joke: If you believe in telekinesis, please raise my hand.

Three types of UFO close encounters were described by ufologist Josef Allen Hynek in his 1972 book *The UFO Experience: A Scientific Inquiry*:[8]
  First: Visual sightings of a UFO.
  Second: A physical UFO event such as electrical interference.
  Third: An encounter with an alien or robot created by alien intelligence.
  This categorization was popularized by the 1977 Steven Spielberg movie *Close Encounters of the Third Kind*, a first-contact movie.

Bonus 2: Carbon and Water, a Marriage Contract for Life. Is It Recognized Everywhere?

Life requires water, nutrients, and energy (carbon cycle). Carbon provides humans with both energy and physical structure. If a molecule is devoid of carbon, chemistry considers it inorganic. Consider the movement of carbon through a cycle of photosynthesis and respiration.
  Photosynthesis is the process where plants feed on the mix of solar energy from the sun, carbon dioxide, and water. In return they produce sugar, starch carbohydrates, and (take a deep breath) oxygen. We eat the sugar and starch (or animals that have eaten plants), and we breathe out carbon dioxide. The process repeats.

None of this would be possible without water molecules. H_2O is a solvent that supports life because the hydrogen (negative charge) and oxygen (positive charge) components act as intermediaries for chemical reactions in living bodies. Now go back and reread the fourth point of the Rare Earth Hypothesis.

In the local search for life, astronomers believe they have evidence of water vapor rising from Europa, Jupiter's largest moon. Europa is a little smaller than Earth's moon, but it harbors a big secret. It is estimated to have twice as much water hidden below its icy surface than Earth has.[9] This makes the possibility of finding life on Europa a lot higher.

If the vapor is confirmed, drilling won't be required to retrieve samples. It just got safer to be an astronaut working the Europa beat. To search for evidence of life, the astronaut can simply test the vapor.

Bonus 3: A World Protocol for Extraterrestrial Contact

How do you say "We come in peace" when the very words are an act of war?

—Peter Watts, *Blindsight*

The International Academy of Astronautics, a nongovernment organization dedicated to promoting science and the technology of human space travel and exploration, have devised a "Declaration of Principles Concerning Activities Following the Detection of Extraterrestrial Intelligence." These rules will save our collective rear ends if extraterrestrial signals are ever confirmed. Below is the protocol (slightly paraphrased) pledged by SETI (Search for Extraterrestrial Intelligence).[10]

1. Any individual, public or private research institution, or governmental agency that believes it has detected a signal from or other evidence of extraterrestrial intelligence (the discoverer) should seek to verify that it isn't human made or a natural occurrence before going public.

2. If the signal can't be traced to a human or natural source then prior to making a public announcement that extraterrestrial intelligence has been detected, the discoverer should inform all other observers or research organizations that are parties to this declaration. They need to do their own checking for confirmation. The discoverer tells her national authorities.
3. If the signal is creditable then the discoverer should inform international organizations including the Secretary General of the United Nations in accordance with Article XI of the Treaty on Principles Governing the Activities of States in the Exploration and Use of Outer Space, Including the Moon and Other Bodies.
4. The confirmed detection of extraterrestrial intelligence should be announced to the public promptly. The discoverer should have the privilege of making the first announcement.
5. All relevant data should be made available to the international scientific community.
6. The discovery should be monitored and recorded for further analysis and interpretation. The data should be made available to the international institutions.
7. If the evidence of detection is in the form of electromagnetic signals, the parties to this declaration should seek international agreement to protect the appropriate frequencies.
8. No response to a signal or other evidence of extraterrestrial intelligence should be sent until appropriate international consultations have taken place.
9. If evidence of extraterrestrial intelligence is confirmed then an international committee of scientists and other experts will act as advisors for continual observation going forward of the discovery.

CHAPTER 16
A REALLY LONG-DISTANCE CALL: INTERSTELLAR COMMUNICATION

The single biggest problem in communication is the illusion that it has taken place.

—George Bernard Shaw

Imagine you are an explorer light-years away from Earth when your ion engine goes kaput. You need to send out a distress call. Of course, you will have to wait decades for a response. Darn you, speed of light—the galactic speed limit! This chapter is dedicated to the difficulty of interstellar communication and provides potential—although unlikely—real science solutions for instantaneous chatter.

Distance isn't the only problem. Time adds another snarl. Oh, yeah, special relativity can be a real kick in the rear for intergalactic civilizations. Think of how different the frames of reference are between a slow-moving planet and an accelerating spacecraft. Some pages have built up since you read chapter 1, but remember the bottom line of special relativity: the faster you move through space, the slower you move through time.

A starship and its home planet will observe the other experiencing time differently (they are in different inertial time frames). So the idea of instantaneous messaging makes my mind spin. What if a ship traveling at warp factor four opens a channel to a ship traveling at sub-light speed to coordinate a tea party with the emperor of the Klax? I challenge you to come up with a nonfictional way for the conversation to occur in real time.

A (not scientific) method used in science fiction is the ansible, a device thought up by Ursula K. Le Guin in her 1966 novel *Rocannon's World*. The ansible sends instantaneous short texts even across a galaxy, relativity be damned. Legend has it she came up with this name because it sounded like the word "answerable."[1]

For the Ender series, author Orson Scott Card ratcheted up the ansible's abilities to full-on conversations. Both authors have interesting techno explanations that I will allow you to research on your own.

In the Crystal series and the Brain and Brawn Ship series, Anne McCaffrey's characters use synchronized crystals. Is this scientifically possible in our universe? Or is it science fantasy, an element that is internally consistent but that doesn't obey the known rules of our universe?

Starfleet in the Star Trek universe relies on a subspace communication network for instantaneous communications. I have yet to hear of a mechanism backed by science on that show. If Spock technospoke to Kirk about how sending a signal through subspace to other points in spacetime actually sends signals through unseen dimensions defined by string theory that reenter space wherever they want, then maybe, just maybe, they'd have a cool explanation.

THEORETICAL LONG-DISTANCE CALLS

Since I'm sure you are familiar with the traditional means of communication, I chose a few far-fetched, but scientifically credible, possibilities supported by general relativity (chapter 1), quantum mechanics (chapter 2), and string theory (chapter 3).

1. Gravity

    Yes, it is scientifically possible (or at least not excluded by science) to use gravity as a tool for intergalactic communication, but it would be tremendously difficult. Pulling this off would require at least a type III civilization.

    Everything we know about the cosmos stems from electromagnetic waves such as radio waves, visible light, infrared light, X-rays, and gamma rays. But because those waves encounter interference as they travel across the universe, they can tell only part of the story. Gravitational waves experience no such barriers and are believed to be unaltered as they ripple through spacetime. Thus they can offer a wealth of additional information. Black holes, for example, do not emit light, radio waves, and the like, but they can be studied via gravitational waves.

Think back to the flat sheet we pulled taut in chapter 1. We are going to reuse this sheet to represent a two-dimensional version of spacetime, only this time instead of placing a bowling ball at the center to represent a star or planet, I toss a rock onto the sheet.

The sheet ripples much the same way a lake does after a pebble has been dropped into it. This is analogous to the ripples of *moving* masses within our four-dimensional spacetime. The more massive the moving object is, the larger the waves are. And, similar to waves in water, they grow weaker the farther out they radiate. An advanced civilization might be able to modulate these gravitational waves to send messages.

Let's blend some science fiction into the science to explain how that might be accomplished. At the receiving side, you will need an Uhura-level communications officer who is experienced in interferometry, the method used by astronomers to detect ripples in spacetime. The sender will use their type III technology to *slightly* manipulate the orbits of small planets (or, more ambitiously, stars) to modulate a signal.

Remember that the speed of this type of communication is limited to the speed of gravity, which is equal to the speed of light. Unless you are dealing with an ancient alien message sent thousands of years ago from thousands of light-years away, this setup is limited to communication with nearby solar systems.

2. Quantum communication

Quantum communication is like the Wild West. Only instead of rounding up cattle, particles need to be lassoed and trained before being sent to a customer. In science speak, this means the particles must be entangled and then teleported away.

Quantum teleportation is the instantaneous transference of the quantum properties of one particle to a different particle. This isn't physical teleportation; it's the transfer of particle measurements. Think of it as teleporting the spin of one particle to another. In an experiment conducted at the University of Calgary in Canada, scientists were able to teleport the quantum states of photons over 6.2 kilometers.[2] The results were presented at the International Conference on Quantum Cryptography on September 12, 2016.

I encourage you to review chapter 2's quantum phenomena like fuzziness, uncertainty, and the wave-particle duality, all of which are the same thing, mostly. To understand quantum communications, you must reacquaint yourself with the idea of quantum entanglement. This is when two particles are said to share the same quantum state despite residing in separate locations. A change in the state of one particle instantaneously changes the state of the entangled partner. This "action at a distance" is also referred to as nonlocality.

After the particles are entangled, one of them must be teleported to another location for the system to work. If the two particles have never been entangled locally then teleportation is impossible. Think of two people who have never met, who live in different countries, and no third person is a common friend to both. The odds of a long-distance text between them are very small.

Quantum particles can only travel so far. They will be lost or become scattered along the way, just as photons are lost as they shuttle along optical fibers. Quantum teleportation systems could include repeaters so the particles could travel farther. The repeaters would need quantum memory to store and pass down the entanglement. Satellites could extend the distance of communication because in space, fewer photons are absorbed or scattered.

Theoretically (the bare minimum required for quality hard science fiction), quantum communication allows for instantaneous communication anywhere in the universe. The important element here is that instantaneous implies faster-than-light (FTL) communication. No wonder Einstein got a headache thinking about quantum physics.

Imagine a three-way civilization led by Earth. Here's the science on one possible communication method: Earth Central Communications Center (the ECCC) entangles two different pairs of particles. Particle A is entangled with particle B, and particle C is entangled with particle D.

Now suppose the ECCC sends particle A to a colony in the Antares system and retains particle B. The organization also sends particle C to a space station orbiting the red dwarf Wolf 359 in the constellation Leo and retains particle D. The ECCC then entangles particles B and D. Presto! An interstellar communication system. Of course the ECCC will charge a hefty fee for that data plan. This might be a reason for conflict, and conflict always makes for good science fiction.

Measurement is the key to a functional quantum communication system. The act of measurement on one entangled particle will collapse the wave function of both particles, and the spin of the measured particle will instantaneously make the other particle spin the same way. Spooky. Again, distance does not matter. Now you have a quantum communication system translating spin into language (similar to Morse code).

Here is the cool bit: because the particles remain in quantum superposition (in all possible states) until measured, any attempt at eavesdropping is foiled. If someone tried eavesdropping she would need to make measurements, and this would randomize the entangled particles. Quantum encryption can secure the internet against data theft or tampering. No matter the distance, quantum entangled photons will instantly mirror any disturbances from one to the other. This is an unshakable system without the encryption key.

In 2016, the Chinese Academy of Sciences put the first quantum satellite into orbit.[3] The plan is to use the satellite to transmit quantum keys to two different locations. If successful, someday a string of satellites might provide a quantum internet for communication.

Quantum entanglement is not only for communication

between people or their civilizations; it can also be used for communication between quantum computers. Imagine a civilization using quantum entanglement to store information and pass it between computers anywhere in the galaxy—instantaneously.

3. Gravitons

A graviton, first mentioned in the atomic interlude, is the quantization of gravity. A way of thinking of quantization, as mentioned in chapter 2's bonus material, is to imagine zooming in so close to an object that it appears pixilated. This pixel is the quantized particle of what you were zooming in on. According to this theory, if you zoom in enough, you will find a discrete particle (pixel) of gravity.

This hypothetical particle puts us squarely back in the world of quantum mechanics. The particle is also a wave, and, if we go "all in" on string theory, the wave might be vibrating in extra dimensions. This extradimensional access might explain why gravity appears so weak compared to the strengths of the other universal forces. It dilutes its strength by spreading across dimensions.

Talk about an opportunity. Dimensional leaking makes interdimensional communication possible. If brane dimensions do exist then gravitons might allow communication with other universes. As described in chapter 3, our four-dimensional spacetime might be a single sheet or a brane among many floating in a higher-dimension bulk. Our first contact with an alien intelligence might not even be from our own universe!

4. Neutrino beam communication

A neutrino is the jealous cousin of electrons. Why jealous? These poor fellows were born with a neutral charge, so they don't get to play in the electromagnetic field with all the other family members that all carry a charge. Its main playmate is the weak nuclear force, one of the four known universal forces and a fan favorite of this book's first interlude. Technically, three different types of neutrinos exist. For the sake of simplicity, I will generalize.

So, where do neutrinos come from? An atom with too many protons or too many neutrons (disrupting its relationship with the strong nuclear force) is cranky and unstable. To feel better, the atom must first suffer through a process called beta decay during which a neutron is transformed into a proton, or vice versa. This happens so that atoms can return to their "happy place," the so-called region of stability. The changes in the atom force it to find a new home on the periodic table of elements.

When a neutron quickly changes into a proton, its new positive charge is offset by spitting out a negatively charged electron. This process guarantees a neutral charge similar to that of the original neutron, but (there's usually a "but" in physics) the sum of the masses of the new proton and the new electron are *less* than the mass of the original neutron. The law of conservation must be obeyed, so the difference between the "before" and "after" is the neutrino particle (technically an antineutrino, but remember, I'm generalizing).

Summary: an atom with excess neutrons will experience beta decay (formally called beta minus decay). This is when a (no charge) neutron decays into a (positively charged) proton, (negatively charged) electron, and a (no charge) neutrino. The sum of the mass of the proton plus electron plus neutrino will equal the original mass of the neutron. Through this process, the atom keeps its neutral charge. Since the atom ends up with a new proton, it becomes a different element. Beta decay can also occur from excess protons (formally called beta plus decay).

By the way, as you read this, tens of thousands of neutrinos are harmlessly passing through you.

It might be possible to use neutrinos for interstellar communication. A sharply focused neutrino beam could send pulsed signals at a target. There wouldn't be much (if any) signal interference because, since they lack electric charge, neutrinos don't interact much with matter. On Earth, we have only sent pulse messages about one kilometer. But if aliens had the technology to create high-energy neutrinos, they could be sending us messages right now. So keep your ears, or rather your detectors, open.

PARTING COMMENTS

The sad truth is that long-distance relationships are difficult. Across the vastness of the cosmos, they might be impossible.

CHAPTER 17
AD ASTRA PER ASPERA: A ROUGH ROAD LEADS TO THE STARS

I am a leaf on the wind . . . watch how I soar.[1]

—Hoban "Wash" Washburne, pilot of *Serenity*,
a space vessel from the TV series *Firefly*
and the film *Serenity*

Up, up and away!

—Superman

The word *spaceship* first appeared in an 1894 novel called *A Journey in Other Worlds* by Jacob Astor IV.[2] We waited until 1934 for the word *starship* to make its debut in Frank K. Kelly's story *Star Ship Invincible*.[3] After that, science fiction was not turning back. It was the moon, Mars, Titan, Proxima b, or bust. This is where the science of rocketry comes in.

Science and science fiction have both come a long way since H. G. Wells used a space gun to shoot a bullet ship to the moon in his story *Things to Come*. At least his use of a space gun is more scientific than the spaceship he devised for *The First Men in the Moon*. The spaceship was made out of fictional cavorite, a material discovered by none other than Mr. Cavor, as a means to negate gravity.[4] If only cavorite were real. Many problems of space travel would be solved, and this book would be a lot shorter.

KEEPING IT LOCAL: ATMOSPHERIC TRAVEL

For planetary travel, our aircrafts rely on the air within the atmosphere. I guess that's why they are called aircrafts. Of course, this assumes that the planet has an atmosphere. For atmospheric travel, three modern types of engines zip across the globe: turbojet, ramjet, and scramjet. All have made appearances in real life and in science fiction.

The more streamlined a jet or car or animal is, the faster it will move given a fixed amount of energy. You should also know that the faster you travel it isn't the friction that will get you (although it does exist), but it will be the air compression that will give you a figurative headache. No matter what you use (bike, boat, or jet plane) or who you are (superhero Flash), movement forward forces air to curve to get out of the way. When the movement is too fast, air molecules can't jump aside fast enough. So they bang into each other, heating up and pushing back.

To go faster through the air (or land or water), the key has always been aerodynamics. At a point near the sound barrier, aerodynamics is no longer enough. The air begins to assert more and more drag, causing extreme turbulence. Traditional turbine engines suck in air and use spinning rotors called turbines to pump the air backward, producing thrust. They will not work at such high speeds because the red-hot blades liquefy.

Here, things get interesting. Unlike turbojets, both ramjets and scramjets (supersonic combustion ramjets) can use that piled-up air without a turbine. No moving parts. They blast air through a specially designed tube and fire it from a nozzle at supersonic speed, producing (in the vernacular of *Battlestar Galactica*) a fraking lot of thrust. Ram pressure is how ramjets and scramjets got their inspired names.

Ramjet combustion only works at subsonic speed (the combustion chamber can't handle supersonic velocities). They can *only* fly through an atmosphere at high speeds and rely on booster rockets in the first portion of their flight. The scramjet burns conventional fuel to reach an initial flight speed before the system starts up. After that it can fly at supersonic speeds because the combustion chamber is designed for supersonic airflow.

THE LONG JOURNEY: SPACE TRAVEL

> *The dinosaurs became extinct because they didn't have a space program. And if we become extinct because we don't have a space program, it'll serve us right!*
>
> —Larry Niven

It is now time to leave Earth and go into space. Travel between planets takes a whole different type of technology than atmospheric flights. Just to get off the planet, you need a lot of thrust. Weight matters. A lot. In good science fiction, weight is always a concern.

To escape Earth's gravity, you need to accelerate from the launchpad to 11 km per second (24,606 miles per hour).[5] So to get off Earth these days, we fire rockets. Another way we could, if we had the full technology, is to use a cable space elevator. The idea for such a system can be traced to 1960 when Russian scientist Yuri N. Artsutanov, in an interview for *Komsomolskaya Pravda*, suggested a cable tethered to an orbiting geostationary satellite (a satellite that remains stationary to a fixed point on the earth) with a counterweight extending out into space above the satellite and the other end of the cable lowered and anchored to the earth.[6] The competing forces of gravity, in theory, result in a taut stationary cable.

If such a system could be made to work, elevator vehicles could carry equipment up into orbit where you build your spacecraft, eliminating the need for large chemical fuel tanks. Lower weight and less fuel? It's a rocket engineer's dream.

Although Arthur C. Clarke didn't think up the idea, he popularized the space elevator to the general public in his novel *The Fountains of Paradise*.[7] Today it is a commonly used tool in science fiction. Someday, I hope it will be common on our nonfiction earth.

Without further ado, here are ways for us to do some space travel.

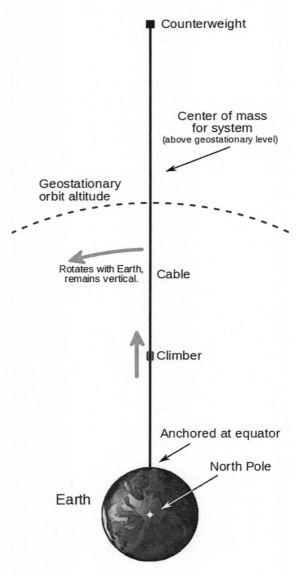

Fig. 17.1. Space elevator diagram. (Wikimedia Creative Commons, author: Skyway, user: Booyabazooka. Licensed under CC BY-SA 1.0.)

Chemical Rockets

All types of chemical engines have a combustion chamber where fuel and an oxidizer are combined. Fuel and oxidizers are nothing more than something that burns and something to start the burning. Three types of chemical rockets are in use: solid, liquid, and hybrid. Each has its own advantages and disadvantages.

Solid propellant engine. The fuel and oxidizer are pre-combined into a solid. Once the burn is activated, the engine won't stop until the fuel has been consumed. The amount of thrust is constant, at least until the fuel is expended. The advantages of this over a liquid propellant are that it offers more thrust and is more stable.

Liquid propellant engine. Both the fuel and the oxidizer are held in liquid form in separate tanks. This type of engine has the advantage of allowing control over the amount of thrust. The pilot (human or AI) can control the mix ratio. The disadvantage is that the oxidizer must be kept extremely cold.

Hybrid propellant engine. Solid fuel is in the chamber while liquid oxidizer is stored in a separate tank. This provides the best of both propellants. It is a very stable fuel with controllable thrust. But it also has some of their disadvantages. It isn't easy to replace the solid fuel (as a liquid would be), and the oxidizer-to-fuel ratio must be constantly rebalanced to maintain efficiency.

Solar Sails

Solar sails use the sun for propulsion. Photons shooting from the sun carry energy and momentum. A solar sail captures the momentum and reflects it off, resulting in continuous acceleration.[8] While a chemical rocket provides a short burst of thrust to speed up quickly, the solar sail starts off with little acceleration that builds up over time. You can only use a solar sail off planet, so it would be cool to have a space elevator.

A groovy electric version has a sail lined with a spiderweb of wires that can be electrified with a positive electric charge—the same charge as the photons shooting from the solar wind. You guessed it: like charges repel and push the ship forward.[9]

Fig. 17.2. Illustration of a solar sail. (Wikimedia Creative Commons, author: Kevin Gill from Nashua, NH, United States. Licensed under CC BY-SA 2.0.)

Nuclear Propulsion

Chemical engines don't have enough thrust to travel far, and solar sails are good for long distances but need time to accelerate. A nuclear rocket wouldn't suffer from either problem. It uses liquid hydrogen that is heated, rather than ignited, in a nuclear reactor. Boom! (The sound effect is to help you imagine a ship powered by nuclear explosions.)

The advantage is the higher amount of energy relative to the fuel for chemical engines. Nuclear rocket technology is popular in near-future science fiction. The *Discovery One* ship from Arthur C. Clarke's original story *2001: A Space Odyssey* used one. The movie version did not.

In Dan Simmons's novels *Illium* and *Olympos*, inhabitants of the far future build nuclear ships to replicate twenty-first-century technology.[10] (As to why, go read the books.) In the television miniseries *Ascension*, the colony ship is Orion-class. (Project Orion, active between 1958 and 1963, studied the use of this type of engine.)

Nuclear rockets do have a problem. They are illegal. The Partial Test Ban Treaty of 1963 (which ended Project Orion) and the Comprehensive Nuclear-Test Ban Treaty of 1996 prohibit aboveground nuclear detonations. Even when launching a rocket from the ground or a low Earth orbit for a peaceful space project, an electromagnetic pulse (EMP) could damage satellites. So this type of rocket would need to be launched away from Earth.

Ion Engine

Ion propulsion is a technology that involves ionizing a gas to propel a craft. Instead of using standard propulsion chemicals, xenon gas (which is like neon or helium only heavier) is given an electrical charge to ionize it. The gas is then electrically accelerated to a speed of about 30 km/second. When xenon ions are emitted at such high speed as exhaust, they push the spacecraft in the opposite direction.

Antimatter Engine

As described in the first interlude, when a particle and its antiparticle meet, their masses are converted into a pair of gamma ray photons with the amount of energy given by Einstein's famous equation $E=mc^2$. Put simply: when they meet, they annihilate each other. This fact gives science fiction plausibility for matter/antimatter engines.

With antimatter, a lot of power comes from only a little bit. A study in 2003 showed that with seventeen grams of antimatter, a spacecraft could cross one light-year of space in just a decade.[11] There are four problems I can think of that spoil this plan. And all of them have to do with one of my least favorite words—practicality.

For starters, it takes a lot more energy to create antimatter than you can ever get out of it. Then there is the cost. Today, we spend about 62.5 trillion dollars per ounce, which feeds into the next problem—availability.[12] Less than twenty nanograms of antimatter have ever been made by humans.

All the antimatter ever made and annihilated at the Conseil Européen pour la Recherche Nucléaire (the European Organization for Nuclear Research, better known as CERN) is only enough to light a single light bulb ... for a few minutes.[13] The antimatter at CERN is mainly used to study the laws of nature. Reasonable fictional solutions include one I used in a short story. The characters used a particle accelerator that circled an entire planet to easily produce antimatter.

The final problem is storage. The property that makes antimatter a great source of energy is the same property that makes it tricky to contain. If you try storing it in anything made up of matter, boom. Instead of attempting to store your antimatter in materials made of normal matter, it might be possible to contain it inside magnetic fields. The science is solid, but the application is difficult.

These hurdles are mountains for scientists but are mere molehills in science fiction. That said, for some reason, I'm still willing to suspend my disbelief on matter/antimatter drives. I'm looking at you, Star Trek universe.

Electromagnetic Drive

This type of engine converts electrical energy into thrust. For launches from a planet, the huge advantage is the lighter weight because all that heavy fuel is unnecessary. This type of engine uses a magnetron to push microwaves into a resonating cavity (a fancy name for a closed cone). The microwaves bounce back and forth, pushing on the cone's wall. More shoving takes place at the narrow end, which pushes the spacecraft forward.

Not all scientists are convinced that this is possible. One group believes this is as effective as moving your car by pushing the steering wheel, or lifting yourself off the ground with a self-inflicted super wedgie. Other scientists think it might be possible if the microwave field pushes against virtual particles popping in and out of the quantum vacuum.

Warp Drive

The inspiration for a warp drive comes from the geometry of general relativity. Did you forget about this? You can find the basics in chapter 1. Spacetime can be warped in a way that will propel a spaceship faster than the speed of light.

Don't worry, we can do this theoretically without breaking any of those pesky laws of physics. As I hope you recall, according to general relativity, the speed limit only applies to objects traveling through space and not space *itself*. Think back to the metaphoric taut sheet with the dimples caused by objects that lie on it. A warp drive would expand a dimple by adding energy.

Quick reminder (mass/energy equivalency): More energy→ more mass→ more mass pressing into spacetime→ bigger dimple→ more gravity.

The drive would push the sheet behind the starship and create a hill (repulsive antigravity). The starship makes the spacetime in front of it dip downward (creating a gravity well). All that is left is for the ship to slide along. Unfortunately, we can't use this type of drive right now because of the amount of energy necessary to cat scratch spacetime.

There is a speculative solution to the energy problem for the warp drive I described—use negative mass to change the geometry of spacetime. Scientists don't know if negative mass exists, but a lot of math has been done to show its properties. Assuming it does exist and we can manipulate this type of exotic matter, we could expand space behind the starship and contract it in front. The starship would then be able to ride along a wave of flat spacetime called a warp bubble. This type of warp drive was proposed by theoretical physicist Miguel Alcubierre in 1994.[14]

GRAVITY AND THE SPACE TRAVELER

Now that you have a taste for space travel, know that astronauts, future space colonists, and people living on the International Space Station need to worry about gravity.

As described in chapter 12, our bodies evolved within Earth's gravitational pull (1g). The prolonged effects of living in a low-gravity environment include muscle damage, bone damage (similar to osteoporosis), loss of red blood cells, and a whole lot more.

A short-term solution for the physical discomfort of space travel is artificial gravity. The long-term answer is adaptation through transhumanism, but for this chapter I'll stick with humans staying as we are today. Isn't this true for most of the characters in science fiction? Here is a list of scientific ways to produce artificial gravity:

1. Take advantage of one of Einstein's greatest discoveries: the gravity-acceleration equivalency. Your starships could constantly accelerate. This is obviously not practical because you would have to constantly accelerate at 1g.

    Acceleration (increasing velocity), not constant velocity, makes you feel gravity. For this to work in science fiction, a starship would need to be orientated to ninety degrees of the direction of thrust. Imagine this orientation for the USS *Enterprise*.

2. Store a lot of mass on your starship. According to general relativity, more mass means more gravity. If you want to maintain 1g, you need to store something that has the same mass as the earth. Science fiction usually has some kind of gravity plate or generator that converts miscellaneous fuels into synthetic gravity.

    Samuel R. Delany came up with an interesting explanation for gravity plates in the novel *Triton*. Einstein's special relativity shows that, as an object is accelerated, its mass increases. At our earthly speeds the increase is negligible, but near the speed of light the effects are (literally) massive. Delany had the atomic nuclei of the gravity plate spin in place at near light speed. Even if we forget about all the power necessary for such a device to work, the plate would have the mass of the earth. Good luck pushing that ship through space.

3. By spinning various sections of your starship, your crew could take advantage of centrifugal force. This worked well enough for USS *Discovery One* in *2001: A Space Odyssey* and the *Hermes* spacecraft in *The Martian*.

Fig. 17.3. Illustration of a spinning wheel and artificial gravity.

4. Your spacecraft could use a gravitomagnetic field generator. I'm including this even though it's more speculative than science. The idea for the generator was first suggested by Russian engineer Eugene Podkletnov in the 1990s.[15] This device hypothetically produces a gravitational field from the angular accelerations of a spinning superconductor.

IF SCIENCE FICTION (ADVENTURE) USES ARTIFICIAL GRAVITY, THEN WHY NOT . . .

In science fiction, starships zip all over the place. Sometimes I don't see (or read) about these ships rotating, or continually accelerating, or having a gravity plate, yet somehow they simulate gravity. Come on! Give me something.

Anyway, no matter how it happens in science fiction, it is done. What I can't excuse are authors who sometimes forget the technology they've created for their world. So, if a space station or starship has artificial gravity, why can't it be increased or turned off? If I'm ever a captain of a starship being boarded by a hoard of space pirates, I would turn up the artificial gravity in certain cabins just as the pirates walked into them. End of story.

A HOPEFULLY TRUE STORY: A TALE OF APPLICATION

First the bad news: given the vast time scale of space travel, we humans most likely will not make it out of our solar system, at least anytime soon. The good news is that this isn't the end of short-term space projects. They just won't involve human bodies. I'm talking about sending technology.

In 2016, Yuri Milner and Stephen Hawking announced an initiative called Breakthrough Starshot.[16] They intend to develop high-speed nanocrafts (sometimes called *nano-spacecrafts* or *nanoprobes*) and send them to the Alpha Centauri star system. Once there, the nanocrafts will relay all types of information including photographs of Proxima b, our nearest earthlike neighbor.

If these tiny crafts can be limited to around a gram in weight then, given the physics of fuel to weight travel, these little guys could be accelerated to a significant proportion of the speed of light. The fuel of choice would be light itself.

A nanocraft could be a protective graphite shell equipped with a microprocessor, radio transceiver, and navigation gyroscope. The tiny craft would then be tethered to lightweight, highly reflective solar sails measuring only a few meters in area. Of course, the sails would be designed to absorb just enough light to get the job done. We wouldn't want them to burn up the craft that they are carrying.

Now, before these nanocrafts can make their interstellar journey, we need to somehow get them outside of our atmosphere (I'm assuming they are built here on Earth). We could launch them with conventional rockets and deploy them in space. Boring. I'd love to see them sent up in a space elevator and launched from a space station, but maybe that's just me.

However they achieve orbit, a ground-based laser array would target the sails and accelerate them to speeds about 20 percent of the speed of light. Once at speed, the solar sails would fold up to become antennas. During the approximate twenty-five-year voyage to Proxima b, the nano-fleet could perform additional scientific tasks. Some could coast past Jupiter's moon Enceladus and sample the water plumes. Others could zip to Pluto (the flight would only take three days).

Only four years after they arrive at Proxima b, we would start receiving messages from our intrepid space explorers. All in all, not too shabby a cosmic project . . . and one that can be completed within a human lifespan.

PARTING COMMENTS

The sky (space) is the limit. Actually, the limit is the speed of light, but I don't want to ruin the moment, so continue looking at stars.

It might not be geology, but aerospace science rocks. We can travel inside our atmosphere in jets (turbo/ram/scram) or, if you like the retro thing, a prop plane. If your intent is to travel between planets, chemical rockets, solar sails, nuclear propulsion (but not launched from the earth), and ion engines are your ticket out of here. If we move from the practical to the theoretical, you could imagine flying starships equipped with antimatter engines, or electromagnetic drives, or warp drives (each less feasible than the last).

THIRD INTERLUDE
A MATTER OF SUBSTANCE

> *What is your substance, whereof are you made, that millions of strange shadows on you tend?*
>
> —William Shakespeare, "Sonnet 53"

Just in case you aren't sick of hearing it yet, here is a reminder: gravity is a force that attracts objects with mass to one another. I'm going to climb a mountain (actually a molehill) to drop some Newtonian mechanics on you.

According to Newton's law of gravity, the attraction between the objects is directly proportional to their respective masses and inversely proportional to the distance between them. If you are standing on Earth, the downward force on you is 9.8 meters/second^2 (32 feet/sec^2). On the moon, the force is one-sixth of what it is on Earth. This is because the size (mass) of the moon is smaller than Earth's.

So, if you weigh 150 pounds on Earth, your weight on the moon is a very svelte twenty-five pounds. On Jupiter, which is eleven times larger than Earth, your weight would be 355 pounds. The good news is that at each location your mass has not changed. The bad news is that it sounds like your weight fluctuates a lot. The force pressing down on you is different when you are on planets with a different mass than Earth.

For completeness, the whole distance part of Newton's law gravity follows an inverse square rule. If the distance between a spacecraft and Earth is doubled, the force of attraction between them falls to a quarter of what it was at the closer distance. Inversely proportional means that when one rises, the other decreases.

SO, WHAT IS MASS?

I hope you are convinced that mass isn't weight. Mass is the actual stuff contained in your body. Mass is also everything around us that is made up of atoms. Everything you see or breathe is made up of stuff. How do we measure mass? It is tricky, after all, to count all the protons inside an object. There is a much (relatively speaking) simpler answer. All you need is to have the object wave at you.

Nope, not a joke. That's what scientists aboard the International Space Station (ISS) have objects do when they want to measure its mass. About 205 miles above your head, the ISS has a contraption that suspends objects by springs on a tray. This, cleverly enough, is called a mass-on-spring oscillator.

Why the oscillator part? Because the mass of the object is measured by determining the rate at which it oscillates. The greater the mass of an object, the slower its oscillation rate will be. This allows scientists to measure mass without all the weighty baggage of gravity.

WHY DO WE HAVE MASS?

Your body is made up of atoms that are made up of electrons, quarks, and a lot of empty space. Since electrons and quarks have very little mass (and space has none), why do you? If you are hungry for more information about quarks and electrons, give the first interlude a read (or a reread). The interlude includes a must-read description of the four universal forces, one of which is, quite shockingly, called electromagnetism.

This is important. Your electromagnetic field is what allows you to *believe* that you are solid. The Standard Model, our current best theory of *almost* everything, includes the Higgs field. This energy field permeates all of space. Envision it as thick honey spread throughout the cosmos.

As a fundamental particle enters the field (this process is called an *interaction*), the particle gets stickier and heavier as it struggles to pass through. This sticky heaviness translates into mass. But not all particles have the same experience. Neutrinos, for example, pass through the Higgs honey without getting too messy. Therefore, they have virtually no mass to show for their journey. Light doesn't care for the taste of honey, so light photons snub the field entirely.

The Higgs field was theorized in 1964 and finally observed by experiment in 2013 at the Large Hadron Collider (LHC) at CERN.[1] (By the way, the LHC is amazing because to date it is the most powerful human-made accelerator ever built.) Only when atoms interact with the Higgs field do they obtain mass.

WHAT MAKES DARK MATTER AND DARK ENERGY SO . . . UM . . . DARK?

> *If you only knew the power of the Dark Side.*
>
> —Darth Vader, from
> *Star Wars: Episode V—The Empire Strikes Back*

Ordinary matter, which is composed of atoms with quarks and electrons, makes up 4 percent of the universe.[2] This type of matter makes up you *plus* all the matter of anything you interact with. What about the other 96 percent? Unseen, but it exists. Twenty-three percent of it is dark matter. The remaining 73 percent is known as dark energy. Scientists might not completely understand dark matter and dark energy, but their existence answers a couple of the big cosmological questions. How did our universe form galaxies? Why is the speed of the universe's expansion increasing?

Dark matter cannot be observed directly because it emits no light, defying electromagnetism. And, as far as scientists currently know, its only friend is gravity. Without dark matter, the amount of gravity hanging around would have been insufficient for galaxies to form.

In the early days, gravity acted as a scaffold around which the structure of the universe developed. Dense, clumpy regions of dark matter created gravitational dimples into which more and more normal matter flowed as the universe aged. This normal material formed stars, which then coalesced to form galaxies. Thank you, dark matter.

Dark energy is the unseen force that tries to resist dark matter's natural desire to clump together over large distances. This repulsive force is built into the fabric of spacetime (those who believe space can be quantized, meaning it is not continuous, will say the *quantum vacuum*). It is therefore evenly distributed and not clumpy like dark matter.

Dark energy and dark matter are competitors in a tug-of-war. Early on, dark matter was winning. As the universe expanded, dark energy became the gorilla on the rope, and the rate of expansion sped up.

IS THERE EVIDENCE OF DARK MATTER?

The gravitational interactions of dark matter with normal visible matter are irrefutable evidence for its existence. This is indirect evidence, which is fine and nice, but scientists sometimes like things to be direct, especially if the theory is to be consistent with the Standard Model of physics. Scientists are working diligently on testable theories. To date, none of their proposed candidate particles have been detected.

The leading contender to explain what dark matter may be is a type of particle called WIMPs (weakly interacting massive particles). WIMPs are thought to only interact with ordinary matter only via gravity and the weak nuclear force. WIMPs are a natural extension of the Standard Model of physics. The model predicts that they were created shortly after the big bang.

Another hypothetical particle to explain dark matter is called the axion. This electrically neutral particle *could* produce the amount of matter currently inferred. These particles are far smaller than WIMPs and were originally predicted to solve a complicated problem associated with the strong nuclear force. As happenstance would have it, it also has the correct properties to be a good candidate for dark matter.

A different answer to dark matter might be sterile neutrinos. Neutrinos, found in the Standard Model, are created after atomic decay. Sterile neutrinos are large hypothetical versions. They are predicted to have no electric charge, so they wouldn't be able to interact with normal matter, but they would be able to interact with gravity. Their effect on gravity would make it appear as if extra unseen matter existed in the universe.

All three particle types can be tested. That's science! However, at the time of this writing, physicists have yet to find any of these particles.

SOMETIMES SUBSTANCES LIKE TO CHANGE THEIR OUTFITS BEFORE GOING OUT

In grade school, students are traditionally taught the three states of matter: solid, liquid, and gas. If you see a grade schooler, you can say, "You are missing two. There are five states (or phases) of matter." I have listed them below. The boundary between these states depends on how active the molecules are.

1. Solid. This state occurs when particles cozy up close to each other. They are very snug with little movement, meaning they exhibit very little kinetic energy (energy from motion).
2. Liquid. This is the state when the particles say to each other, "Whoa, back off a bit, but stay in the friend zone." This arrangement keeps them close enough to flow around each other but not close enough to form shapes. Because this state holds some distance between the particles, there is more room for them to dance around. Ergo, more kinetic energy.
3. Gas. This is the state where the particles realize that they aren't real friends but just acquaintances. They stand far enough away from each other to express themselves actively (free-style dance) and have a lot of kinetic energy.
4. Plasma. This state is like that cousin who rarely comes to visit. You know the particles are out there, but they don't come into contact with you often. Particles in this state are highly charged, which makes them jump about a lot. They have extreme kinetic energy. If you use electricity to ionize neon gas (or any of the noble gasses), it phases into plasma and glows. Our sun is in a plasma state.

5. Bose-Einstein condensates (BECs). This type of matter occurs when the particles come very near to stopping all movement and have almost no kinetic energy. The atoms begin to clump together and form superatoms. Superatoms arise when groups of atoms get clingy and form a single atom. In an experiment conducted with BECs, the speed of photons (light) were slowed.[3]

Speaking of Bose-Einstein condensates, let's get weird but stay real. Imagine an artificial quantum matter supersolid. I can't. My mind doesn't work that way. I have a few mental tricks for quantum mechanics but this, not so much. A supersolid is rigid like crystal but simultaneously acts as a superfluid, a fluid that flows without friction.

This is a real thing. Using lasers, two separate teams (the Swiss Federal Institute of Technology[4] in Zurich and MIT[5] in the United States) of scientists nudged the quantum state of a BEC, which is a superfluid, to behave simultaneously like a solid.

CHAPTER 18
WHY ARE WE SO MATERIALISTIC?

The hand is the tool of tools.

—Aristotle

Just about anything your great-great-grandparents made for themselves or bought was mostly formed from wood, cotton, wool, brick, and iron. Few of the elements found on the modern periodic table of elements were used in production a century ago. Think of an element as an ingredient used by nature and chefs (chemists) to make materials. Today we use these ingredients to make a lot of new and exotic materials that would blow a nineteenth-century engineer's mind.

Of course, we could have done that with dynamite. Nitroglycerin did exist back then, but I was speaking figuratively. Our creation of ceramics, plastics, semiconductors, fiberglass (glass made up of glass fiber), metamaterials, and a lot of other wild stuff described in this chapter just didn't exist until very recently. Nowadays we also make stuff using ingredients that *aren't* found on the current periodic table of elements.

WHAT IS MATERIAL ENGINEERING?

Material engineering is the creation of materials. It has existed since *Australopithecus garhi* flaked off bits from rocks to form crude tools about three million years ago. A millennium later, the Stone Age really started hopping. It had a good long run before petering out about ten thousand years ago. During this period, the key materials for tools were stone, bone, and animal skin. Copper, tin, and gold were all known during the Stone Age, but they were predominantly used ornamentally because they were too soft for use as tools.

The Bronze Age kicked off in China about 1,700 BCE. At that point, someone discovered that copper and tin could be combined to form bronze, a very strong metal resistant to rust. Iron was common enough, but it ranked low in popularity. It was not much harder than bronze, but it was difficult to melt. You can melt bronze in a pot while you need a furnace system to work with iron. So the people of this era decided, why bother?

Only some did bother. And because of them, everything changed when they stumbled upon a new material. It turns out if you add just a dash of carbon to iron, steel is formed. The discovery probably happened accidently by repeatedly putting iron back into the fire to work on it. Each time it returned to the coals, a bit more carbon was added.

Because steel is stronger than bronze, less of it is needed to forge sturdy tools or weapons, making them a lot lighter. Steel also holds an edge better than bronze, a perk for weapons and tools. This Iron Age began in 1,200 BCE in the Middle East and spread to southeastern Europe and China around 600 BCE. Bronze was still widely used for objects like statues and fountains because it resisted rust better than objects made of iron. These three ages can overlap quite a bit depending on the location.

ARE ANY NEW MATERIALS UNDER DEVELOPMENT NOW?

The discovery and use of new materials is occurring faster today than during any other time in history. Today's material engineers are always searching for ways to create new materials with specialized characteristics. For example, a material imaginatively named SAM2X5-630 has the characteristics of glass but is stronger than steel.[1]

If this sounds a lot like the fictional transparent aluminum container used to contain the Earth-saving whales in the movie *Star Trek*

IV: The Voyage Home, then you get the idea. There is a lot of cool potential for this material. Not only is it a glass-like metal, but it is also very elastic. Satellites made from this material would be able to deflect meteors. Possibly of more interest to you, your smartphone would be nearly indestructible and would bounce when you dropped it.

Google is working on developing smart contact lenses that will use flexible electronics to display a diabetic's glucose levels.[2] Soon silicon will be replaced with cheap, flexible materials for solar cells and computer displays. Want to make an athlete (perhaps yourself) happy? Offer her smart clothing such as running shorts with a tiny embedded accelerometer to sense movement and adapt to a runner's stride. Or a shirt that helps a player's tennis swing.

SPEAKING OF ATHLETES, HERE ARE A COUPLE OF (LITERALLY) COOL PROOFS OF CONCEPT

Thanks to those pesky laws of thermodynamics, anytime we cool air, we create heat through the energy used to create cooling. Turn on an air conditioner to cool your bedroom, and the motor heats up. This is also true of material science. Unsure what the laws of thermodynamics are all about? You won't be (I hope) after you read their formal description in chapter 21.

If you wear a cotton shirt while being chased by Daleks down endless hallways, your heart pounds away and there is a good chance that your body is overheating. Your shirt absorbs the radiation emitted from your body (which cools you down) and stores the heat. An alternative would be to wear a high-tech shirt that uses your own sweat to keep you cool.

Researchers at Stanford University are studying a new material called nanoporous polyethylene,[3] which, unlike cotton, lets the heat radiation escape. Additionally, unlike the high-tech wicking shirt you might have worn on your last ten-kilometer run, it doesn't need sweat to keep you cool.

Nanoporous polyethylene achieves its goal when the external temperature is lower than your body's. If this material is sandwiched between cotton cloth pocked with tiny holes for airflow and treated with a chemical to wick away water, you get the perfect workout outfit. Researchers are studying how flexible it can be made and how it works on human skin.

Don't worry, material engineers haven't forgotten their roots. It makes perfect sense for them to look to the animal kingdom for inspiration, just as their Stone Age ancestors did. Small animals such as beavers evolved fur that traps air. This air provides a layer of insulation whenever the animal enters water.

Now the spacing between hairs and the length of the hairs on artificial pelts keep the air level intact. In water, it maintains the air layer. Scientists are modeling small animal pelts so they can create swim gear for cold-water divers.[4]

IT'S ELEMENTARY

An element is something made of a single atom. For example, hydrogen is an atom made up of one proton. If you add a proton (think fusion) to hydrogen, it becomes a helium atom once. The elements have been good to us inside and out. Your body and everything you can ever interact with are made up of elements.

Scientists have done a lot with the known ingredients they've found, all of which are listed in the periodic table of elements. After the addition of nihonium, moscovium, tennessine, and oganesson in 2016, the tally stands at 118.[5] Granted, these newbies are stable for only a fraction of a second and therefore not very practical for making new materials, but they do add to our understanding of the earliest moments after the big bang when these higher-order elements might have existed.

Even with the 118 that have been found, there might not be enough elements on the periodic table to meet the demands (or dreams) material engineers have for the next generation of synthetic materials. The solution is simple. Just make a bigger table with more elements. Add a little bit to nature. This is where superatoms come in.

Superatoms occur when a group of regular atoms are combined to act like a single element, one not found on the periodic table. A

WHY ARE WE SO MATERIALISTIC? 241

superatom can have properties that don't exist when the mundane elements are combined. Chemists are working out the rules for linking them into molecules.[6] Once they do, they'll be able to create synthetic structures that can be tweaked to serve specific purposes in exotic materials.

These superatoms might be why Tony Stark can make his Iron Man uniform. Otherwise, how could it be so light and flexible? At face value, it defies the properties of non-synthetic elements. Iron and titanium are too heavy for his movements or for flight.

As long as we are talking about members of the Avengers, Captain America's shield is made from vibranium, a metal/material I challenge you to find the elements for. In the *Avengers* movie, the shield is somehow able to displace the kinetic energy of blows from Thor's hammer (whatever that is made of). Following the rules of physics, my math kung fu cannot begin to work out the displacement that makes that possible. Unless Thor's hammer isn't all that powerful. And yet, it is.

WHAT IS A SUPERCONDUCTOR?

Superconductors are a type of material in which electrons are liberated and therefore free to flow between atoms without resistance. And as a perk, no heat or stray energy is released from their motion. On a superconducting material, an electric current could flow endlessly without needing a power source to give it a swift kick in the rear to keep it moving.

It sounds like the perfect material, and it would be if it didn't have one tiny (almost not worth mentioning) problem. The electrons are only free when the material is unenergetic, meaning it has to be very, very cold, near absolute zero cold (-273 degrees Celsius/460 Fahrenheit).[7] It takes a lot of energy to cool a material enough to make it superconductive. In *Star Trek*, if Spock were an economist, he might tell Kirk that the cost of superconducting outweighs the benefits.

The holy grail for material scientists is to discover a material that acts as a superconductor at room temperature.

ANY WAY AROUND THAT FREEZING PROBLEM FOR SUPERCONDUCTORS?

The answer is a work in progress. Perhaps the solution will be found with the most basic of the elements, hydrogen. Rival teams of scientists are racing to turn the simplest, most common, and first among all elements in our universe into a metallic room-temperature superconducting material.[8]

The record for warmest superconductivity is held by hydrogen sulfate at -70 degrees Celsius.[9] I bet you knew that hydrogen sulfate contains hydrogen. If hydrogen can successfully be made into a metallic material, it could replace copper in electrical wiring, reducing the loss of energy that comes with copper. Best of all, that efficiency would decrease the demand for power.

As is true for most matter, depending on temperature and pressure, hydrogen comes in different flavors or, rather, different phases called states. The third interlude includes details for the different states of matter. At room temperature and earthly atmospheric pressure, hydrogen is a gas. As pressure increases, it will transition into a solid. Mix in more heat to have it transition from a solid into a liquid. Under extreme pressure, the electrons begin to dislodge. The hydrogen changes into *either* a liquid or solid metal. If it becomes a solid and is able to maintain that state after the pressure is released, we can have our superconducting wiring.

A CARBON SOLUTION FOR MATERIALS

Carbon is the sexy hero element of the day and not just because it's the element that makes you organic. When it pulls off its glasses, it goes from mild-mannered element to the superhero of exotic materials. One is graphene. It's such a fantastic material that Andre Geim and Kostya Novoselvo won the 2010 Nobel Prize in Physics for isolating the material in 2004 at the University of Manchester.[10]

Graphene is the thinnest of carbon materials, the width of a single carbon atom, and it is flexible. That's not all. Additional benefits are that it's stronger than steel, flexible, transparent, and unreactive. Although it might not be a superconductor, it is the most conductive of the current materials. Due to its slenderness and flexibility, it is the material of choice for nanotechnology.

At the Tsinghua University in China, researchers discovered that spiking the mulberry leaves they feed silkworms with a graphene solution makes the little critters spin silk that is 50 percent stronger.[11] This super silk also has the ability to conduct electricity. This new material could someday be used to make wearable electronics.

Carbon knows another cool trick. If we arrange the atoms a bit differently, we form a carbon-based material called Q-carbon. It gets the "Q" designation because forming the material involves heating and rapidly cooling the atoms in a process called quenching. Besides the cool name, it is both magnetic and stronger than a diamond. And, believe it or not, it glows in the dark![12]

This material is perfect for many applications like electronic displays or abrasive coatings for biomedical sensors. Not that it matters much, but the Q can be used as a quick way to make diamonds. Carbon has many structural forms (graphene, graphite, nanotubes, buckyballs, diamonds, etc.), but none are magnetic. This property makes the Q-carbon structure so exciting.

By the way, a nanotube is rolled-up graphene, and buckyballs are hollow balls made of carbon atoms that can be used as medical delivery systems.

CAN ANY MATERIALS HARVEST LIGHT?

A polymer-based photovoltaic material can be used to harness not just the sun's visible light but also the infrared rays.[13] The new plastic material uses quantum dots mixed with a conducting polymer to absorb infrared. Quantum dots are tiny semiconductor particles made up of cadmium telluride that react to electricity or light. The switch from silicon to plastic solar cells could make it easier and cheaper to soak up the sun's rays for energy.

Previously, plastic solar cells relied solely on visible light. They were only able to convert an anemic 6 percent of the sun's energy into electricity.[14] Using this new material, plastic solar cells might be able to achieve up to 30 percent efficiency.[15] Want more? Okay. This material can also be used to detect infrared light for night vision cameras. The current cameras rely on semiconductor crystals. Imagine swapping out expensive crystals with plastic.

NOW YOU SAW IT, AND NOW YOU DIDN'T

What do you think about people seeing through you (optically, that is)? Or, how about cloaking your starship as you sneak past an enemy armada? If you have thought about these things, then read on. But beware: the methods of invisibility I'm going to describe are more Klingon than Harry Potter. This is invisibility of the science kind.

Two scientific methods can render an object invisible: spatial cloaking (making something invisible in space) and temporal cloaking (making something invisible in time). These theories are straightforward and easy to understand; however, the methods of their application are very complex. In science fiction, the opposite is usually true: the theories are based on a lot of fantastical elements of the novel's world, and application is simple—we push a button, or drink a formula, or wave a magic wand that is sometimes called a sonic screwdriver, and presto chango! We're invisible.

The key to invisibility is light manipulation. As light reflects off an object, it carries information about the object (such as its size and shape) to our eyes or to mechanical detectors. So, the goal of invisibility is to prevent tattletale light from passing on these details.

1. Spatial Cloaking

    Using this nifty trick, light flows around an object and recombines on the other side, leaving no evidence of its detour. If you were to look at the object, you would only see what's on the other side. The physics behind this isn't much different than how refraction makes a straw appear bent in water.

    As usual, the difficulty (not so difficult in science fiction) is in the *how* of building the cloak. It all comes down to the

WHY ARE WE SO MATERIALISTIC? 245

construction of metamaterials, materials that have properties not found in nature. In spatial cloaking, optical metamaterials made from nanotechnology surround the object to bend light.

Think back to your last vacation when you were getting your Zen on. You might have watched water flowing evenly around a stone in a river. This is what metamaterials do. They flow light around an object. Of course, the low-tech option uses mirrors to direct light around what you are trying to hide. Although this is magician simple, it isn't as efficient because it's a stationary form of invisibility. With a metamaterial cloak, you (might) be able to walk around unseen.

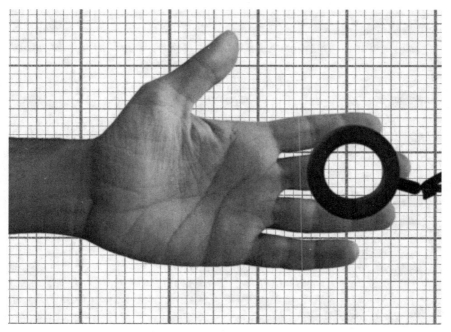

Fig. 18.1. Illustration of spatial cloaking.

2. Temporal Cloaking

Instead of cloaking an object, why not cloak an event? Whenever light illuminates an event, it marks that event in time. If an event somehow managed to leave light unperturbed, then the event would remain hidden. How can this be done? Easily, when you exploit a characteristic of light: speed.

General relativity shows that time can be slowed, or at least its appearance relative to a different frame of reference can be slowed. A time lens takes advantage of this time-bending property. As light passes through a time lens, it will split; that is, the light scatters. Some of the light will speed up, and some will slow down as the lens changes their frequency and wavelength, creating a zone of darkness.

These separate pieces of light pass through a second lens that recombines the streams. Any event that occurred inside that dark gap won't have scattered any light. It will appear as if it never happened. There will be nothing to be seen. This is a time gap. This is not science fiction. In a study that appeared in the journal *Nature*, a group of Cornell University scientists created a time gap forty trillionths of a second long.[16] Yeah, I know. Not much time. Still, it is proof of concept.

In science fiction, a nefarious villain could hold the technology needed to lengthen the time gap. Then, during the gap, the fiend could attempt to commit the perfect crime, one that appeared to never have happened. (Except for the criminal consequences, of course.)

No matter how it ends up being used, for now, the science of temporal cloaking offers science fiction a lot of cool phrases like manipulating the perception of time, time lens, and time gap, all of which are actual concepts in real science.

Now for the difficult question: how can we get it to work? Once again, this isn't as difficult for a science fiction writer as it would be for a physicist. Temporal cloaking requires special lasers and optical fiber metamaterials to disperse and reconstitute light. We might need a few more years for this one.

PARTING COMMENTS

The title of this chapter asks, why are we so materialistic? The answer is, because we have to be. After all, things need to be made of things.

Warfare, clothing, and food production are the great motivators for material discovery. The different types of materials humans have used over the centuries include stone, wood, bronze, iron, steel, cotton, and wool.

Material science emerged during the nineteenth century as the study of the structure of the elements within substances. Modern material engineers are developing new materials, for products that function better, and new technologies. This includes metamaterials, materials with properties beyond what is available in naturally occurring materials.

One such property might be invisibility. Two scientific ways to render an object invisible are spatial cloaking, which disguises an object in space; and temporal cloaking, which disguises an event in time.

CHAPTER 19
TECHNOLOGY (COOL TOYS)

We tend to overestimate the effect of a technology in the short run and underestimate the effect in the long run.

—Roy Amara, researcher and futurist

We wanted flying cars, instead we got 140 characters.

—Peter Thiel, entrepreneur

1. Anything that is in the world when you're born is normal and ordinary and is just a natural part of the way the world works.
2. Anything that's invented between when you're fifteen and thirty-five is new and exciting and revolutionary and you can probably get a career in it.
3. Anything invented after you're thirty-five is against the natural order of things.

—Douglas Adams, *The Salmon of Doubt*

This chapter is gadget-driven. It is completely free from mind-staggering questions of existence or galactic origins. Here it's just cool technology. We might not have flying cars, but as you've seen from the examples throughout this book, we aren't doing too shabby. Yes, instead of flying cars, the wish of the 1950s, we got the internet. I can live with that.

THE LASER

Before becoming a plaything of scientists and either tools or weapons in endless science fiction stories, it was an acronym: Light Amplification by Stimulated Emission of Radiation. A laser spits out energetic coherent photons (quantum light particles) focused on a tight spot.

Think of it as light amplified via a crystal and focused rather than scattered. Remember that light is also a wave, and the term *coherent* is a fancy way to say that all the peaks and valleys of light waves are lined up. This also means they are all the same color. If they were solid, they could be nicely stacked on top of each other. This is called being in-phase.

The first lasers were built in the 1960s. They are used in computers (in optical disk drives), printers, surgery, cutting tools, tools to measure distances between objects in carpentry, and really destructive weapons in science fiction.

POPULAR NAMES FOR LIGHT-ENERGY WEAPONS IN SCIENCE FICTION

Laser (go with what you know)
Heat ray (*The War of the Worlds*)
Ray gun (used in early science fiction, generally before lasers were invented)
Death Ray (the 1930s' Nikola Tesla idea for directed energy weapons)
Phaser (short for phased array pulsed energy projectile weapon in the Star Trek universe)
Pulse rifle (science fiction outside of the Star Trek universe)
Blaster (*Star Wars*)
Lasgun (*Warhammer 40,000*)
Plasma gun (not a laser; learn all about plasma in the third interlude to find out why)

3-D PRINTING

Although not quite a Federation replicator from the Star Trek universe, the nonfiction 3-D printer comes close. It is capable of printing out physical objects from an inputted digital model. The printing is an additive process that creates copies by laying down successive layers of material.

> I'll let you
> I'll let you know
> I'll let you know when I have
> I'll let you know when I have finished
> I'll let you know when I have finished making a
> I'll let you know when I have finished making a 3-D-printing joke.

If you have the necessary materials, the only limit on what you can print is your imagination. Of course, you need the correct materials to load your printer. I'm pretty sure plastic polymers might not make the best mug of beer.

One of the most fascinating and practical things about 3D printing is medical applications. Print up some organs for drug testing. No more mice needed. All you have to do is take a scan of the patient, which creates instructions (a digital model) for the printer. These instructions are sent to the printer's nozzles, which spew out a gel-like mixture composed of mature tissue cells, stem cells, and polymers designed to mimic real tissue consistency. The organ is printed in a layered lattice, leavings channels throughout that will act as blood vessels so that nutrients can circulate.[1]

3-D-printed body parts have been implanted in test animals.[2] The Integrated Tissue and Organ Printing (ITOP) system begins by printing layers of biodegradable materials into the form of the desired tissue. This scaffolding is then filled with gels containing living cells that develop into functional tissue. Implanted materials in the scaffolding can encourage bone growth so that the scaffold can be removed.

Although it will be a while before personalized organs are available for transplants, biologists are getting closer. In 2001, the human bladder became the first bioprinted organ successfully implanted into a human.[3]

And now for a feel-good proof of concept. In 2011, when Kaiba Gionfriddo was six weeks old, he started turning blue. He was having difficulty breathing because one of his bronchial tubes had collapsed. In early 2012, a 3-D-printed windpipe was used to hold the airway open.[4] It dissolved a few years later, allowing his bronchus time to grow strong enough for normal breathing.

HOW ABOUT A TECHNOLOGY COMBO MEAL WITH A 3-D PRINTER?

Here is a combination of the tech ideas (all real) described in this book that we can use to create some solid science fiction . . . that might actually become our future. It all starts with a 3-D printer, and it ends with colonization.

First our engineers, using some of the rocketry technology from chapter 17, transport 3-D printers to a marginally inhabitable exoplanet, perhaps Proxima b (chapter 12). Once planet-side, robots (chapter 14) are rolled out to use the printer technology to create a habitable base for future human colonists.

GREAT TOOL, BUT ARE THERE ETHICAL ISSUES WITH 3-D PRINTING?

I'm sure you can think up a lot of medical issues, but there are a few that might not be as obvious, at least not right away. Such as making a gun in your back den, or a neighbor producing cocaine, or a terrorist manufacturing ricin. Sorry for bringing you down. If you like to dress up, the next section will hopefully cheer you back up.

WEARABLE TECHNOLOGY

Wearable technology is so much easier than having to lug around equipment. It is also much sneakier if you have nefarious intentions. I'm sure you have seen this in science fiction. Yes, I know the Fitbit is great for your health, and augmented reality glasses like Google Glass dumping information on you is all great and interesting. But I want to describe wearable things you might not know about such as the Bluetooth dress by Sony Ericsson. This cocktail dress lights up when the wearer gets an incoming call.[5] Is it going to change the world? I don't know, but it is fun.

How cool would a computer tattoo be? The technology is getting close. Scientists at the University of Tokyo have built a prototype of a computer screen that can be worn on your skin.[6] The two-micrometers-thick polymer LED is equipped with organic photodetectors connected to a sensor. The whole contraption is attached to the volunteer with material very similar to plastic food wrap. Its job is to measure blood oxygen levels. It doesn't sound like much right now, but think of it as the first step in a journey that will eventually attach your e-mail to your wrist.

If you aren't into tattoos, how about a bracelet? There is one that can analyze chemicals from a drop of sweat. Then it sends the data to your smartphone.[7] Someday it might even be able to detect molecules linked to depression.

CAMERA TECHNOLOGY

Imagine how cool it would be to have a camera that takes photographs around corners. You might actually be able to buy such a camera in your lifetime. A proof of concept device already exists at the Heriot-Watt University in Edinburgh where researchers modeled bats using a form of echolocation to build a specialized camera.[8] Similar to how a bat may use sound (echolocation) to map its surroundings, these researchers bounced laser light off the floor and scattered it in all directions. Their camera was able to detect the light echoes as it smacked a test object around a corner.

COMPUTER IDENTIFICATION

Computers that recognize who you are as soon as you walk into a room. I'm not talking about the weak AI of your web browser that has collected information so it can flash a banner ad for products you might like, or your social media platform of choice predicting the political party you favor from what you post and who you follow. I'm talking about computers that know you for you.

It is very nearly possible with today's technology that when you walk into your smart home, the Wi-Fi system will identify you. A human body partially blocks the radio waves between router and computer. Members of a family tend to come in different shapes and sizes, and they even walk differently. This builds a pattern that is sent to the family computer via Wi-Fi.

As of this writing, scientists have built an algorithm that is 95 percent accurate at distinguishing two adults and 89 percent accurate with six adults in the same room.[9] A system like that could identify you and tailor heat and light settings to your preference without asking. Mix in a bit of science fiction and it could detect your emotional state and cue up appropriate music, perhaps even mix a drink for you after a particularly trying day negotiating trade agreements with one of the outer colonies.

MACHINE MIND CONTROL

> *Today's best brain-computer interface systems are like two supercomputers trying to talk to each other using an old 300-baud modem.*
>
> —Phillip Alvelda, manager for DARPA's Neural Engineering System Design program

The Defense Advanced Research Projects Agency (DARPA) wants humans to be able to control machines with their minds. This is not such a far-fetched idea for either good fiction or reality. Their wetware project, called Neural Engineering System Design (NESD),[10] implants a small disc into the brain. This device is measured in millimeters and converts the brain's chemical signals into a digital pulse.

This idea isn't particularly new, but NESD can connect up to a million individual neurons, which makes it easy to *think* your electronic tools into being more productive. You wouldn't have to leave the couch. Best of all, people with prosthetic limbs would have a lot more control.

PARTING COMMENTS

> *The guy who invented the first wheel was an idiot. The guy who invented the other three, he was the genius.*
>
> —Sid Caesar

CHAPTER 20
WHAT DOES IT TAKE TO EXIST?

Reality is that which, when you stop believing in it, doesn't go away.

—Philip K. Dick

Do you think you are real? Of course you do. What if I told you that you are really a hologram projected from quantum fluctuations at the boundary of the universe? Would you still believe you are real? My answer, for what it's worth, is yes! *Cogito ergo sum.*

By the way, what does *real* even mean? I suppose reality can be defined as our brains' interpretation of the messages sent up from our five senses. Science is prefaced on the assumption that the external world exists and has properties that can be observed. We use our sensory data (or robotic senses) to draw in data, and we use our brains (or computers) to interpret that data.

Our brains create three-dimensional images from two-dimensional data. Obviously, this questions the existence of things beyond our senses. For example, did atoms only exist after we developed the technology to detect these tiny building blocks? Or is it possible that atoms always existed, senses be darned? I don't think our belief in atoms is what made them real. I'm reasonably sure they existed before the idea of an atomic model came to Democritus in 400 BCE or when we finally saw them using an atomic resolution microscope in 1983.

Then again, thinking about something, as Democritus did with atoms, does not necessarily mean it will ever be proven to exist. I'm sure some people believe in unicorns, and yet I have never found one in any zoo or free-ranging around my garden. We have to be careful to not get circular and claim that our consciousness thinks up the idea of atoms and we are composed of atoms so that we can have a brain to have consciousness and think up atoms. I'm going to use the definition of reality to mean that something can exist without us needing to believe in it.

DO ANY OF OUR IDEAS ABOUT REALITY MATTER?

> *Reality leaves a lot to the imagination.*
>
> —John Lennon

All that is necessary for science is that (pick the theory of your choice and write it here) it agrees with observations and that predictions can be made. We might never know the true nature of the reality of our universe. Maybe the anthropic principle (see chapter 5) is correct, and we live in the habitable region of the multiverse where the rules of physics give rise to life, and different rules apply to other regions so that each region might have its own reality. But that isn't science. It's philosophy.

As an example of conflicting realities, I'll describe the parable of the goldfish that Stephen Hawking and Leonard Mlodinow used in their book *The Grand Design*.[1] Imagine a goldfish hanging out in her fishbowl. The curved sides of the bowl as well as the water give the fish a much different view of reality than we enjoy. The fish sees the path light bending as it hits the water. We landlubbers, meanwhile, see light travel a straight path.

If the fish is intelligent enough to work out the physics of light from observation (as we mostly have), she would be able to formulate scientific laws and make predictions based on the curved light she sees. Because of the curvature and aquatic effects, the fish's scientific laws will be more complicated than ours. But is hers any less valid a perception of reality than ours? After all, she can make observations and predictions as accurately as we can. Is the fish's version of reality wrong and ours right?

What if, with all our crazy complicated (but correct) ideas of quantum mechanics, we are the ones inside the fishbowl?

WHAT DOES IT TAKE TO EXIST? 259

We might not be mentally responsible for the creation of the physical world, but we can exert some thought control over it. We can augment it in ways that provide a deeper understanding of our surroundings. Or we can forget trying to understand it and create an alternate virtual reality that is completely under our control. An interesting theory covered in this chapter proposes that our entire universe might be a virtual reality. To paraphrase Shakespeare, all the universe is virtual, and we are but holograms.

WHAT IS AUGMENTED REALITY?

Augmented reality (AR) is an experience of reality supplemented by computer input, usually provided through displays embedded in mobile devices such as smartphones or eyewear. What you see through your tech window is developer content blended with the real world to provide information about what you see. It pushes the limits of photorealism.

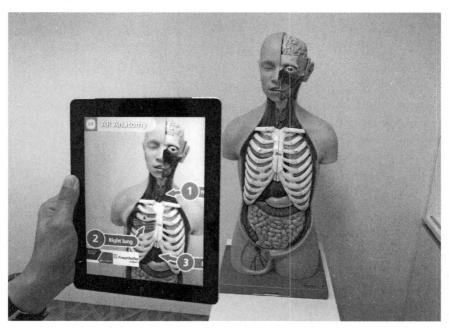

Fig. 20.1. Illustration of augmented reality.

A lot of market applications utilize AR, such as gaming and advertising. In 2016, people all over the world burned up their smartphones' battery times in their quest for Pokémon. The game *Pokémon GO* encouraged a scavenger hunt for Pokémon characters while players learned facts about local landmarks. They held up the phone and used it as a window to see both what was really there and what had been digitally added. In *Pokémon GO*, you saw your target (augmented) hanging out in the natural environment.

Before you get frustrated during your next trip to a foreign land because you can't read a sign that directs you to the nearest subway, simply use your smartphone's camera to allow an app to translate it for you. It worked nicely for the Doctor in the *Doctor Who* episode, "Under the Lake." In that show, the twelfth version of the Doctor uses his sonic sunglasses as a language translation scanner.[2]

WHAT IS VIRTUAL REALITY?

Oh, the places you'll go!

—Dr. Seuss, *Oh, the Places You'll Go!*

Virtual reality (VR) creates an illusion of being somewhere you are not. The "virtual" in VR means it's all an illusion. The worlds we create using this technology can be so much bigger than the physical world in which we live. The created space can be bigger than the Milky Way even though the device that creates the space is the size of an Xbox. The only limitation on the size of our creation is human imagination.

Now that I think about it (and I bet you have also), I would enjoy having my illusions brought directly to me on my own personal holodeck. Federation crew members in *Star Trek: The Next Generation* frequent the holodeck for fun and fancy. Except for the "solidness" of the holodeck fantasies, the technology for the VR experience more or less exists today.

WHAT DOES IT TAKE TO EXIST? 261

WHEN DID THE SCIENCE OF VIRTUAL REALITY BEGIN?

In the early 1880s, scientists figured out how the brain processes the separate 2-D images that are viewed by each eye to render a single mental 3-D image. Devices that took advantage of this picture-processing trickery such as the stereoscopic viewer, including the popular View-Master (an ancestor of yours probably used one), provided a sense of immersion. In 1929, the first flight simulator was created.[3] Instead of pictures, the simulator was controlled by motors that mimicked turbulence and physical feedback resulting from actions taken with the controls.

Fig. 20.2. Illustration of stereoscopic vision device.

The next major advance in brain trickery came in 1961 with the first motion-tracking head-mounted display.[4] These fancy, oversized goggles provided stereoscopic visual effects along with stereo sound.

That's not all. This headset included a magnetic motion-tracking system connected to a camera. Simple head turns moved the remote camera, allowing the user to look around wherever the camera was located. This contraption was called Headsight and was created for remote viewing by the military.

In 1969, the practical idea of artificial reality was developed. It was much more complex than the simple remote viewing, and computer-generated environments in which multiple users could interact were created. People miles apart could communicate with each other in a computer-generated environment.

Sound familiar? As in, something you use today? Or something you've seen in the 1999 movie *The Matrix*, where characters live in a simulated world completely unaware that they aren't living in the *real* physical world? I'm not saying it could happen, but you might want to review artificial intelligence in chapter 13.

Two big things happened in 1987. First, the term *virtual reality* was finally used.[5] Second, haptic technology was developed. Haptic technology is clothing (such as a glove) that can mimic the sense of touch during a simulation. The same glove could translate the user's motions into the simulation.

Today you can be immersed in a virtual world using any of the image-filled goggles on the market. In some VR games, you can wear specially designed clothing that provide tactile experiences that enhance the images floating around in your head. If you choose to go off market, you could make your own VR device. All you need for a headset is a high-resolution screen with a wide field of vision and a fast-reacting screen that refreshes images in real time as your head moves.

ANY PRACTICAL USES FOR VR UNRELATED TO THE MILITARY AND ENTERTAINMENT?

Today's VR devices are used for a lot more than first-person shooting games or active (ahem) participation in porn. They are used in medicine, aviation, tourism, and psychological therapy.

The 1992 movie *The Lawnmower Man* was an early fictional

attempt to show how VR might be used for therapy. Jobe, a mentally challenged greenskeeper, is introduced to VR environments that (unscientifically) improve his intelligence. Somehow using the computer connectivity of the early 1990s, he gains power over all linked devices. And because it is a movie, mayhem ensues.

VR exposure therapy is a real thing. It is being studied as a treatment to relieve symptoms of post-traumatic stress disorder (PTSD).[6] Patients confront their traumatic memories through virtual exposure. This type of virtual exposure can also be used to treat phobias.

Exposure therapy in the form of storytelling can be used to foster empathy toward others ... which is perfect for VR stories. For example, *Perspective; Chapter 1: The Party*, created by Morris May and Rose Troche in 2015, allows a viewer wearing goggles to experience sexual assault at a frat party from the man's and woman's perspective.[7] Perhaps this type of storytelling can be incorporated into college orientation week.

CAN FALSE MEMORIES BE CONSIDERED VIRTUAL REALITY?

False memories represent something that did not really happen, and yet a person can feel as if he lived through it. Philip K. Dick wrote a short story for *The Magazine of Fantasy & Science Fiction* in 1966. "We Can Remember It for You Wholesale" is about a clerk for the West Coast Emigration Bureau who is bored with his life. He is unable to afford his dream of visiting Mars, so he opts for the cheaper alternative of having memories of him having been there implanted ... only to discover that he really had been to Mars.

He follows the clues in his past as more real memories return. Like many of Philip K. Dick's stories, this is a wild trip into what reality actually might be. The story was adapted to film in 1990 as *Total Recall* starring Arnold Schwarzenegger. A remake of *Total Recall* starring Colin Farrell was made in 2012; that version keeps the story terrestrial by dropping the Mars angle.

Imagination plays a big role in the creation of a false memory. In a cautionary study out of Stanford University, researchers were able to show how false memories can be prompted in children by a third party using VR technology.[8] This technology could be used in marketing, or perhaps someday false memories will provide children with skills (from supposedly already knowing it) quicker than traditional instruction.

WHAT ABOUT ERASING MEMORIES?

Instead of false memories being added, what if original memories could be removed? How would that affect a person's sense of reality? The movie *Eternal Sunshine of the Spotless Mind* plays on the idea that memories can be erased; in this case, the erasure is attempted after a romantic relationship ends.

The story has both ethical quandaries and brain science that isn't too far off. The science to end a memory exists, but being able to selectively choose which memory to zap is still science fiction. Unless the *Men in Black* really do have a mind-wiping magic wand. Below are a couple of drug treatments currently being studied to erase memories.

1. After a compound called ZIP (short for zeta-pseudosubstrate inhibitory peptide) was injected into the hippocampus of a rat, an earlier fear of being shocked while on a slow-spinning carousel experiment was erased. In the experiment, each time the rat rotated past a particular point, the critter was shocked. It learned to turn around and head in the opposite direction. After the injection, it forgot about the shock and enjoyed the carousel ride.[9]

    Long-term memory is often considered the process of neurons firing simultaneously (or nearly simultaneously). The activity develops a bond that will likely make them fire together in the future. For example, the neuron that has encoded a particular scent can be associated with the neuron that causes you to tear up.

    The ZIP drug blocks communication between these complementary neurons. The problem of using ZIP-like drugs

is the targeting of individual memories. In humans, no biomarkers separate out good from bad memories.
2. Where short-term memory ends and long-term memory begins is difficult to quantify. Do you remember this book's introduction? I'm thinking that I shouldn't bet on it. I am sure you know what you had for breakfast, if you ate one. Unless you have the same thing every morning, you probably won't remember this morning's breakfast a year from now. If the introduction had an emotional impact, though, then it went into your long-term memory.

Something that affects you emotionally, either pleasantly or not, releases norepinephrine, a promoter of a protein cocktail in the amygdala (the processing center for your emotions and the fight/flight/freeze response to a threat). If a drug that lowers the production of norepinephrine could be applied shortly after a traumatic event, it might prevent the formation of a long-term memory about the trauma. Medics on a battlefield could use the drugs with soldiers to cut down on instances of PTSD.

HOLOGRAPHIC REALITY

Is being called two-dimensional an insult? It shouldn't be, because what we perceive as our three-dimensional universe might be nothing more than two-dimensional source code projected across a horizon. Some speculation proposes that we might all be holograms, scattered light that is reconstructed into a 3-D representation of the original 2-D version. This theory is called the holographic principle.

The holographic universe was first proposed by Gerard 't Hooft of the University of Utrecht in the Netherlands.[10] If it is true, are you real? Or rather, the type of real you *think* you are? The principle suggests that you are actually a 3-D representation of a flat 2-D version of yourself that is hanging out at the boundary of the universe (wherever that is).

If we really are all holograms, a lot of cosmological questions would be answered, including how to connect relativity with quantum mechanics. The principle suggests that gravity can be described in a theory in one less dimension that has no gravity. A 2-D universe without gravity is able to project the effects of gravity (and even black holes) in a 3-D universe.

Isn't this a bit like the stereoscopic vision we talked about, where we perceive 3-D from 2-D images?

LET'S GET DEEP INTO SOME SCIENCE FICTION

I have described a hologram as reconstructed light, but I left out an important bit of information: the light can be reconstructed at some future time long after the original source is gone. The holographic principle provides science fiction creators with another solid idea that can be useful: time travel.

Star Trek: The Next Generation came close to this idea, only instead of holograms, they used teleporter technology. In the episode "Relics," Captain Montgomery Scott (famously known as Commander Scotty, the chief engineer for Kirk's *Enterprise* in the original *Star Trek* series) is reconstructed in the twenty-fourth century after having his pattern stored for seventy-five years in a teleporter buffer.[11]

Another science fiction idea is to completely give in to the holographic principle and accept that earthlings reside in a holographic universe. Then perhaps a select few heroes and villains become self-aware enough to modify the coding of their 2-D templates. What would happen? Would they be godlike in their abilities? Or, perhaps by simply shifting the position of a hologram, a starship would have the appearance of flying faster than light. All of these wild ideas are consistent with the holographic principle.

The next time you read in DC Comics about all the reality-warping power of Mr. Mxyzptlk (from the fifth dimension), just know that he might get these powers by manipulating a 2-D projection of the earth from a higher dimension. He does love a good practical joke at the expense of poor old Superman.

HOLOGRAPHIC THEORY AND BLACK HOLES

The holographic principle also suggests that the event horizon of a black hole is a recorder and a projector. It records all the information of what has fallen into it. According to this model, nothing ever really falls into a black hole. Rather, objects that enter become spread out around the event horizon. The interior of the black hole is nothing more than an illusion and therefore inaccessible to anyone on the outside. As the black hole evaporates, all the stored information is encoded in the radiation. Therefore this principle offers up a theoretical solution to the black hole information paradox (a contradiction between general relativity and quantum mechanics) described in bonus 2 of chapter 7.

PARTING COMMENTS

> *Reality is frequently inaccurate.*
> —Douglas Adams, *The Restaurant at the End of the Universe*

Your view of the world (your reality) can be computer augmented with real-time information. If you are less interested in the reality you are stuck with, change it. Technologies to create user-generated realities exist for leisure or therapy. Keep in mind it might be that messing with reality is a slippery slope. Hello, *The Matrix*. The same is true for adding false memories or erasing true ones.

Now that you know about the holographic principle, I want to leave you with something to consider. If *Star Trek: The Next Generation* had been written to exist in a holographic universe, it would have been turtles all the way down (meaning infinite regression) because on the holodeck, you'd be creating holograms within holograms.

CHAPTER 21
THE END OF EVERYTHING

The End Is Nigh.

—Walter Kovacs, aka Rorschach,
from Alan Moore's *Watchmen* (1986)

This isn't the happiest of chapters. Fittingly placed at the end of this book, the chapter is all about endings. As long as time has a direction, then atoms, humans, the earth, the sun, the Milky Way, and the universe will all at some point . . . just end. Everything has an expiration date. This final chapter explains the how and why of some of the grand finales.

SPEAKING OF ENDINGS (AN AUTHOR'S CONFESSION)

I remember one time when I was standing by the side of the road holding up a sign that read, *The end is near! Turn around and save yourself!* This rude driver shot me the finger as he passed, yelling that I was an apocalyptic a-hole.

To be honest, it sort of hurt my feelings. Anyway, next I heard tires screeching and a loud splash. Then it hit me. My sign should probably have said, *Bridge out.* Oh well, live and learn, or die.

Okay, back to the science.

THE END OF THE INDIVIDUAL

All life, all the way down to cells, has a metabolism, and all metabolisms wear down. Chapter 9 covered fraying telomeres and the Hayflick limit along with other aging indicators. So, evidence of you getting older does exist. For some people, looking into a mirror just isn't enough. Chapter 9 also gave some scientific hope for biological immortality, and then chapter 10 hinted at downloading your mind into a computer.

It isn't just us carbon-based organics who wear out. Whether they are our friends or overlords, machines need to be considered in this ending game. If it has moveable parts, it ages over time and ages more quickly with use (wear and tear).

If you think we can all go on forever as posthumans who share our virtual living space with AIs, then you need to know one key fact: the party could go on for quite a while, just not forever. Thanks a lot, entropy! I will tell you all about the "e" word soon. For now, just know that both metabolic and mechanical systems need fuel. Someday the universe will be unable to provide it.

As you continue along the path of this chapter, you will discover that all the trails toward immortality fade away. The word *immortality* should include "really, really long but temporary" in its definition.

THE END OF THE HUMAN SPECIES

Something to ponder: our species might be the first to notice an upcoming extinction event ahead of time.

Unlike the end of an individual, the end of humanity is very speculative. Many different statistical models predict the number of years we might have left. Some include extinction-level events like asteroid strikes, a nuclear war, and overpopulation, all of which chum up the science fiction waters.

Others are based on the average number of years a hominid species has survived in the past. In a speech given at Oxford University's debating society, Stephen Hawking stated his belief that, unless we get off the earth, our species has about one thousand years before extinction.¹

This is more optimistic than the fifty-fifty chance of surviving until the end of this century astronomer Martin Rees proposed in his gloomy book *Our Final Hour: A Scientist's Warning—How Terror, Error, and Environmental Disaster Threaten Humankind's Future in This Century—On Earth and Beyond*.² His probability calculation is based on the danger of us not truly understanding how destructive our technology really is. I wonder if he reads a lot of science fiction.

I'm not going to give you any odds based on my beliefs. Instead I'll present some facts and you can draw your own conclusions. Let's begin with a few of the local endings from our past. Today it is estimated that 99.9 percent of all species that have ever existed are gone.³ The culprits behind the extinctions are volcanic eruptions, asteroid impacts, and temperature change. These are natural causes. The human-made ones were discussed in chapter 11.

The greatest "hits" of Earth extinctions:

Ordovician-Silurian 440 million years ago (MYA), 60 percent to 70 percent of all species went extinct. The probable cause was a combination of falling sea levels and glaciation.⁴

Late Devonian 360 MYA, 70 percent of all species went extinct. The probable cause was global cooling and depleted oxygen in the oceans.

Permian-Triassic (also known as the Great Dying) 250 MYA, 96 percent of all marine species and 70 percent of land species were extinct. The cause might have been the acidification of oceans from dissolved carbon dioxide thanks to volcanic eruptions. So much for the bulk of the species coming out during the Cambrian explosion. (See chapter 8 for what exploded.)

Triassic-Jurassic 200 MYA, 70 percent to 75 percent of all species became extinct. The probable cause was the sudden release of carbon dioxide from volcanic activity.

Cretaceous-Paleogene (also known as the K-T extinction) 65 MYA, 75 percent of all species became extinct. Bye-bye, dinosaurs. The possible cause could have been an asteroid impact, or volcanism, or a combination of the two.

272   BLOCKBUSTER SCIENCE

The Holocene extinction (nicknamed the "sixth extinction" and also referred to as the Anthropocene extinction): this is happening now. Until the industrial revolution came along, two vertebrate species out of ten thousand went extinct every one hundred years. According to that number, nine species should have disappeared in the past century. Instead, 477 disappeared.[5] If the increase in greenhouse gas continues at its current rate, one in six species will face extinction by 2099.

ROID RAGE (DESTRUCTION BY ASTEROID)

Fig. 21.1. Illustration of an asteroid hurtling toward Earth.
(iStock Photo/RomoloTavani.)

Big rocky chunks are inconsiderate enough to crash down on Earth about every 500,000 years. This includes asteroids and their speedy comet cousins. The goodish news is that these rude drop-ins haven't all been extinction-level events. The badish news is that, according to the geological record, an asteroid extinction event has occurred every twenty-six million years.[6]

Scientists are seeking an explanation for this pattern. Here are some of their ideas for why the earth gets stoned, none of which are related to peer pressure. They are all speculative and extraterrestrial.

1. Planet X

    A Neptune-sized planet might be orbiting the sun every fifteen thousand years and periodically causing comet disturbances. Although not directly observed (if only it were that easy), its theorized existence is inferred from the gravitational tugs on stellar things we can see, most of which are things we don't want tugged. Hence our interest.

    Astronomers estimate that Planet X has ten times the mass of the earth, so if we knew where to look, we'd probably be able to see it with a powerful telescope. For now, we can only theorize its orbit.[7]

2. Companion star to the sun

    A hypothetical star named Nemesis is postulated to be hanging out past the Oort cloud and circling our sun in a wide, elliptical orbit.[8] With a name like Nemesis, how could it not be evil? As it periodically approaches, it disrupts billions of comets in the Oort cloud with its gravity and shooting them at the earth.

    By the way, Nemesis's nickname is Death Star. The math for its orbit might work out, but there is no evidence for its existence. Remember, math can drive science, but it is not science itself.

    If Nemesis does exist then it is probably a brown dwarf, a small star with insufficient mass to maintain the nuclear fusion of hydrogen. This means it will be small and dim and therefore hard to detect directly.

3. Oscillations through the galactic plane

    Instead of (or in addition to) another sun or planet perturbing the comets, it might be that our own sun's wavy journey through the Milky Way causes all or some of the problems. The sun moves up and down as it circles the center of the Milky Way. Its route is not much different than that of a wooden horse bopping up and down on a carousel.

    This oscillation exposes our solar system to gravity differentials (changes in gravity) that affect material in the Oort cloud. The movement also changes how much cosmic radiation we are exposed to. For example, the sun passes through the north side (upper region) of the Milky Way about every sixty-two million years.[9] Because it is exposed to an excess of cosmic rays during this time, our environment is impacted with weather changes.

For the record, there is no "up" or "north" in space.

Fig. 21.2. Illustration of the merry-go-round Milky Way.

4. Dark matter

This is not about asteroids, but it does provide a possible explanation for the natural pattern of extinctions. As the sun bobs through the galaxy on its carousel, it might pass through clumps of dark matter, an acquaintance of yours from the third interlude. These particles will fly through the center of the earth and possibly affect the core's temperature. The results would be volcanic eruptions (hello, Krakatoa) and rising seas. This is the most speculative of the causes for mass extinction because, other than gravity, scientists aren't sure what dark matter interacts with.

IS ANYONE WATCHING OUT FOR THESE ROCKS AND ICE BALLS?

To detect these wandering asteroids and comets, astronomers rely on both ground-based and space-based telescopes. The Jet Propulsion Laboratory at the California Institute of Technology has recorded about fifteen thousand near-Earth objects (NEOs).[10]

Not all asteroids are created equal. The dinosaur killer of the Cretaceous-Paleogene extinction, which left its footprint in what is now modern-day Mexico, was about ten kilometers wide (6.2 miles).[11] Ninety-five percent of asteroids this size or larger have been identified, and their orbits are tracked by NASA. Ninety percent of medium-sized asteroids, ranging from one to ten kilometers (0.62 to 6.2 miles) wide, are under similar surveillance.

The small ones, less than one kilometer wide, are largely unknown. They can drop down faster than a bullet. Most of them will explode in orbit due to the kinetic energy involved with hitting the atmosphere. Those that survive are most likely to land in oceans. If they were to hit land, cities could be damaged, but they do not pose extinction-level threats.

A twenty-meter meteor, labeled 2012 DA14, exploded twelve miles above Chelyabinsk Oblast in Russia in 2013.[12] The explosion was about thirty times greater than the atomic bomb dropped on Hiroshima. The shock wave spread out and damaged thousands of buildings, but fortunately no deaths were reported.

IF THE BIGGER NEOS TARGET EARTH, CAN THEY BE STOPPED?

Detection is only half of our defense system. The other is deflection. To stop an asteroid, you need to nudge it a bit and alter its path. Early detection is therefore crucial. The farther away the uninvited guest is, the less of a nudge is necessary to change its trajectory. If it is far enough out from Earth, our space engineers might only need to change its velocity by a couple of millimeters per second.

How could they do it? The boring answer is that they hit it with a projectile. A more elegant solution is to use a gravity tractor. This is when you send an object (artificial or otherwise) to a position near the asteroid and allow the nearly infinitesimal gravitational attraction between them to nudge the asteroid slightly.

An offending comet might be treated a little differently, and in a way that is a little showier: launch a rocket and detonate a nuclear bomb close to the ice ball. If the math is calculated correctly, enough of the surface will boil away to change the comet's trajectory so that it bypasses Earth. If this happened in science fiction, the engineer would be sweating out the complicated calculations. Keep in mind that comets travel twice as fast as asteroids, and your team has only one rocket.

No matter your choice of deflection, it will clearly have more chance of being effective than the non-science method used for the 1998 film *Armageddon*. That idea was so zany it isn't worth addressing in a book interested in science. I will, however, comment on the crazy way the heroes get to the asteroid, which involves a slingshot maneuver of a 1990s' NASA space shuttle around the moon and then landing it on the offending rock.

I'm not certain how they failed to notice an asteroid the size of Texas until it was only eighteen days away.

As fate (coincidence, or possibly film studio competition) would have it, the movie *Deep Impact* came out the same year. This one is about a comet hurtling toward Earth, and the plot attempted to include actual science. And it didn't do a bad job. Not perfect, but not bad. Keep in mind that an Orion-class ship is a nuclear rocket (see chapter 17 for details), not one that uses a chemical engine as is mentioned in the movie.

Some social circles (they call themselves Whovians) claim that the asteroid that killed off the dinosaurs was sent by the transhuman Cybermen in an effort to destroy Earth. *Doctor Who* companion Adric tragically sacrifices himself in the attempt to save our planet. It didn't work for the dinos, but it worked out okay for mammals. For Adric haters, it was a complete success. I can nitpick this episode until tomorrow's breakfast, but the science of this *Doctor Who* episode makes more sense than that of *Armageddon*.

THE SUN AND EARTH: A RELATIONSHIP THAT ENDS WHEN THE LIGHTS GO OUT

Our sun is 4.5 billion years old. It is nearing the halfway point of its yellow phase where its energy predominately comes from fusing hydrogen into helium. We like this phase. It has been good for life on Earth. However, you should know one important thing. Although our sun appears yellow through our atmosphere (don't look at it unfiltered!), it is really white. The yellow wavelength travels all the way through the atmosphere while the blue wavelength is scattered (blue skies smilin' at me, and all that). I'll call it a yellow star to avoid confusion.

Now, don't get too comfortable thinking that human civilization has at least another 5.5 billion years before we must exit this planet as the sun goes into its red phase. Over the next two billion years, the sun will expand and grow hotter as it begins to run out of hydrogen to burn. Helium will look tastier and tastier as an energy choice.

The increase in heat that will result will change Earth's carbon cycle. This will lead to more water vapor in the air and runaway greenhouse effects. We will be the new Venus on the block. If humans are still on the planet during this two-billion-year transition, they will hopefully have moved toward the poles or underground. At the very least, we would need to set up the solar shields described in chapter 11 to block out the rays.

In about 5.5 billion years, as the ratio of helium to hydrogen fusion increases (ending its yellow phase), the sun will become red and bloat outward. The Goldilocks habitable zone will be pushed farther and farther outward. The sun will also become lighter. The drop in gravity will cause all the planetary orbits to widen. This is where the physics get really tricky, and by tricky I mean there is no consensus on the amount of mass that will be lost or the degree of orbital changes. Will our red giant sun expand far enough to engulf Earth, or will Earth's expanding orbit save it?

It won't matter. At that point, it won't be habitable for human life. Earth is done with us. The atmosphere and oceans will have evaporated from the increased heat. But we already left, right? I bet that Saturn's moon Enceladus is looking pretty good right now. The temperature on Mars might not be too bad. I hope that before this era arrives, our geoengineers have read chapter 12 and worked on terraforming other locales in the solar system.

The sun has one more set of fireworks to ignite. After about 150 million years performing as "Big Red," it will run out of helium to burn. This will signal its final curtain call. It will shrink until helium from its outer regions pushes down on the core, ignites the outer layer, and blows it into space in clouds of gas and dust. The process will repeat until only the core remains.

The core will continue to collapse for another 500,000 years. Eventually the sun enters its elderly phase and will be known as a white dwarf. No more fusion occurs, but it is still hot. Whatever is left of humanity can steer their asteroid colonies or self-contained VR servers close enough to stay warm and absorb energy to fuel their uploaded minds. This will work until the white dwarf fades to black over another (speculated) trillion years. After that? We had better have moved to other stars.

WHY WILL ONLY SUPERSIZED GALAXIES EXIST IN THE FUTURE?

The Milky Way galaxy is part of a gang made up of fifty-four galaxies. Astronomers came up with the intimidating name Local Group for these thugs. Most of the gang members are small, but the Milky Way is a big player along with the Andromeda and Triangulum galaxies.

The Local Group is gravitationally bound, meaning that the gravity holding the members together is stronger than the universe's expansion. This will be true for a long time but not forever. Gravity will lose in the long run to the stretching (dark energy) of the expansion. As nonmember galaxies move away from Earth, the stars of the Local Group are actually getting closer.

A rumor claims that the gang members don't always get along well. The word on the street is that in about four billion years, the Andromeda and Milky Way galaxies are going to clash.[13] Yes, a collision is imminent. They will first pass through each other (our fading sun will be safe because stars aren't all that close to each other). Then, thanks to gravity, they will snap back toward each other.

This will happen over and over until they combine into a single galaxy. And if the super massive black holes at the core of these galaxies collide as predicted, a lot of energy will be released as gravitational waves. Eventually, this galactic martini shaker will apply mixology between all fifty-four gang members until only a single super-galaxy that I shall now name DAVID is left. This naming is unofficial. I'm not authorized to name this new hybrid galaxy in the science world. But in the universe of this book, I am almighty.

This process will happen to all local groups across the entire universe. All that will be left are super-galaxies separated due to expansion.

WE CAME ALONG AT A GOOD TIME (TO PREDICT THE END OF THE UNIVERSE)

We have evolved in the universe at a very fortunate time. It hasn't grown so big that we can't come up with really cool cosmological models that describe its origin and its future. Future civilizations might not be so lucky.

If an intelligent race of beings evolved in one of the super-galaxies, what could they learn about the universe? More importantly, what could they learn of the DAVID system? Nothing, actually. They will not be able to see any other galaxies once the other super-galaxies have faded over the horizon faster than the speed of light. They won't know about DAVID.

Their observational science will determine that theirs is the only galaxy in the universe. No evidence for an expanding universe or a big bang will exist. A large portion of our universal history is written in the cosmic background radiation (even more chapter 4 goodness), which at some point will have been stretched so thin it will become undetectable. This civilization will conclude that the universe is eternal and unchanging and that they are the only galaxy in it.

THERMODYNAMICS WILL BE THE END OF US, BUT FIRST LEARN ITS LAWS

As the prefix thermos- suggests, thermodynamics is the study of heat, and three laws govern this area. A zeroth law about temperature change kicks it off. Below is a description of the laws followed by an easy summary sentence to help you remember and understand the meaning of each.[14]

0. If the temperature is higher in a system that comes into contact with another system, its temperature will fall while the temperature in the other system will rise until both temperatures are the same.
    Easy way to remember: there is a game and you must play.
1. Energy and matter cannot be created or destroyed within a closed system. This law is called the law of conservation of energy.
    Easy way to remember: you can't win; you can only break even.
2. The entropy of any closed system that is not in a thermal equilibrium will almost always increase. Thermal equilibrium means no temperature differences exist within the system. Entropy is the amount of a system's energy that is unavailable for work.
    Easy way to remember: you can't break even.
3. Entropy in a closed system approaches a constant value as temperature approaches absolute zero. Absolute zero is the lowest temperature possible where movement of molecules has the least amount of kinetic energy (movement).
    Easy way to remember: you can't quit.

Summary of thermodynamics: if you think things are a mess now, just wait.

HOW DOES ENTROPY EXPLAIN THE DIRECTION OF TIME?

From the second law of thermodynamics we derive entropy, the measure of disorder within an isolated system. Consider your bedroom. If you make no effort to clean it, it will gradually become messier. Perhaps quicker for some than others. The only way to prevent the messiness is to regularly clean the room.

In physics, the equivalent is to pump energy into a system. The addition of energy is the only way to slow down or reverse entropy. However, in an isolated system such as our universe, the first law of thermodynamics states that energy *cannot* be added, so entropy (disarray) will continue increasing. If no universal housekeeper pops in for a cleaning, everything everywhere will gradually fall into disarray.

The earth avoids complete disarray because our sun pumps some of its energy into our system and pushes back against entropy. Unfortunately, the sun will eventually run out of energy.

The general rise in entropy gives us a sequence of events physicists call the *arrow of time*. The arrow does not tell us that the past leads to the future. It only tells us that it goes from order to disorder. This is an important distinction. When entropy reaches its maximum value, the arrow of time will break.

HOW IT ALL ENDS

The first law sets us up for the fall. The law of conservation of energy ensures that the amount of mass and energy at the start of the universe remains constant throughout any point in its history or future. Thanks to the expansion, the universe is stretching out a fixed amount of matter. Eventually it will become so thin it will be as if it never existed. Thanks to dark energy, it will expand faster and faster. Matter and energy will thin out until at some point even individual atoms are pulled apart.

The third law of thermodynamics makes the dire prediction of the universe's ultimate fate. In the far distant future as the universe veers toward disorder, there will be no more thermodynamic energy (it will have been stretched too thin). This doesn't mean heat won't be present but that heat differentials will no longer exist. A single universal temperature will rule. This is called the *heat death of the universe*.

WHEN ALL THE SUNS TURN OUT THE LIGHTS

All the stars will eventually run out of fuel. The main sequence stars described in chapter 15 will become white dwarfs. After that, they cool down into black dwarfs that will feed the massive black holes left in the universe. In a nonillion years, all that will be left are black holes. That's pretty far in the future. A nonillion is described in chapter 4.

The black holes will try to feed on the tiniest bits of energy left, but the pickings will be slim, so they will begin to evaporate. Even the evaporated radiation isn't much to talk about. It will be stretched so thin that its energy (wavelength) will be negligible. The universe will get close to the absolute zero described by the third law of thermodynamics, the heat death with a broken arrow.

The universe will be an unchanging, cold place with no energy or mass. There can be no life, mechanical or organic, because that requires atoms and energy. There will not even be enough energy for thought.

LIFE AFTER THIS UNIVERSE

Perhaps before the heat death, if the universe stretched enough, spacetime itself might tear into another dimension into which posthuman life might escape.[15] Or perhaps much earlier, posthumans of a type VI civilization (chapter 6) decided to tour the multiverse (although the multiverse technically isn't science). Now, a type VII civilization might try to pump in energy from the universes they have created outside the universe. This would drive back the entropy.

PARTING COMMENTS

I know. Not a happy chapter. You shouldn't feel too despondent because a bit of good news can be found. Science has not excluded the possibility that some life-forms (or artificial intelligences) might survive until the end of the universe . . . or possibly outlive it!

You might be the person to make it happen. I'm sure you could think up interesting and perhaps unexpected ways to combine the scientific ideas in this book. That's how much confidence I have in you. If you are not a scientist then perhaps you could use your ideas to create great science fiction or a cool video game.

This kind of speculation drives scientists and creators of science fiction into the realm of Arthur C. Clarke's second law of prediction: the only way of discovering the limits of the possible is to venture a little way past them to the impossible.[16]

Bye for now. There is a lot more science and science fiction to share in our universe.

ACKNOWLEDGMENTS

Somewhere in the hypothesized multiverse there may be an Earth where this book doesn't exist. A very sad Earth because we would never have gotten together to celebrate all things science and science fiction. So first up, I want to acknowledge you the reader for being in the correct universe. Thank you.

Quantum mechanics has shown that not even space itself is a pure vacuum. Quantized field interactions are happening everywhere; this is what makes the universe such an interesting place. Yes, science has shown that nature really does abhor a vacuum. In the scaled-up universe, writers and scientists also shy away from working in a vacuum. It takes a team to do good science, and it takes a team to put together a good book. With that in mind, I want to begin with thanking Maryann Karinch, the agent who believed in my project from the beginning.

I would also like to express my gratitude to the staff of Prometheus Books. They are an amazing group of people who helped make *Blockbuster Science* into something special. In particular, Jeffrey Curry, who read the manuscript (I can only guess how many times), offered suggestions, and helped me track down sources. Because of me, I suspect he can't stop from talking techno babble to his friends. Next up, Hanna Etu (we share a love of science fiction), who helped me through the publishing technicalities. And then there is the editor in chief, Steven L. Mitchell. He signed me up, believing in my idea of promoting science by linking it to science fiction. I know I annoyed him with my jokes, but he kept me around anyway. Thanks. I want to acknowledge the readers the Prometheus team found. They challenged me with science questions and which topics to expand on.

A shout-out goes to my writer's group, Words-in-Progress. Trust me, the membership was never timid about telling me what didn't work. Of particular annoyance (I mean helpfulness) were Jim Kempner and Skip Seevers.

I can't forget Laine Cunningham and her editing. She helped me see this project through to publication. Also, I need to shine a light on Leya Brown who helped me track down some of the illustrations.

A special thank-you goes to Susanne Shay, whom I'd go to when I had trouble making a sentence work. She'd always agree: it didn't work. Then she'd pull out her tools and help me fix it.

There is this kid named Seth Bernstein, who not only supported this book but also reminded me that Tony Stark is a cyborg because of his artificial heart. Stark's dressing up in an Iron Man suit to play with the Avengers is more of a personal choice. I made sure I got that correct in the book. Thanks, son. Speaking of my children, I want to thank Gwendolyn for being my number one fan. I know I have a repeat customer with her. I want to thank my wife, Michelle, for putting up with me during this project.

Finally, I want to thank my father, Murray, for watching *Doctor Who* with me when I was a kid and taking me to see *Star Wars*. Because of that, this book is his fault.

NOTES

INTRODUCTION

1. William Wilson, *A Little Earnest Book upon a Great Old Subject* (London: Darton, 1851).

2. The philosopher and theologian William Whewell formally proposed the word "scientist" in 1840 in his work *The Philosophy of the Inductive Sciences*. "We need very much a name to describe a cultivator of science in general. I should incline to call him a Scientist." Whewell was really into nouns. William Whewell, *The Philosophy of the Inductive Sciences* (London: John W. Parker, 1840), p. 113.

3. Arthur C. Clarke, *Report on Planet Three and Other Speculations* (New York: Harper & Row, 1972).

4. Arthur C. Clarke, *Profiles of the Future: An Inquiry into the Limits of the Possible* (New York: Popular Library, 1973).

CHAPTER 1: ONCE UPON A SPACETIME

1. *Venture* 49 (September 1957).

2. Albert Einstein, trans. Robert W. Lawson, *Relativity: The Special and General Theory* (New York: Henry Holt, 1920).

3. David Nield, "Physics Explained: Here's Why the Speed of Light Is the Speed of Light," *Science Alert*, April 13, 2017, https://www.sciencealert.com/why-is-the-speed-of-light-the-speed-of-light (accessed June 19, 2017).

4. Joe Haldeman, *Forever War* (New York: St. Martin's Press, 1974).

5. Orson Scott Card, *Ender's Game* (New York: Tom Doherty Associates, 1991).

6. John Archibald Wheeler and Kenneth Ford, *Geons, Black Holes, and Quantum Foam* (New York: W.W. Norton, 2000), p. 235.

7. Neil Ashby, "Relativity and the Global Positioning System," *Physics Today*, May 2002.

8. U.I. Uggerhøj, R.E. Mikkelsen, and J. Faye, "The Young Centre of the Earth," *European Journal of Physics* 37, no. 3 (April 8, 2016).

9. Tom Siegfried, "Einstein's Genius Changed Science's Perception of Gravity," *Science News*, October 4, 2015, https://www.sciencenews.org/article/einsteins-genius-changed-sciences-perception-gravity (accessed 6/30/2017).

10. A. Einstein and N. Rosen, "The Particle Problem in the General Theory of Relativity," *Physical Review* 48, no. 1 (July 1, 1935): 73–77.

11. Paige Daniels, "The Outpost," in *Brave New Girls: Tales of Girls and Gadgets*, eds. Paige Daniels and Mary Fan (CreateSpace Independent Publishing Platform, 2015).

12. Madeleine L'Engle, *A Wrinkle in Time* (New York: Farrar, Straus & Giroux, 1962).

13. S. W. Hawking, *Black Holes from Cosmic Strings* (Cambridge: Cambridge University Press, 1987), p. 5. Published in *Physics Letters*, B231 (1989): 237.

14. *Star Trek: The Next Generation*, "The Loss," first broadcast December 29, 1990, directed by Chip Chalmers and written by Hilary Bader.

15. US Energy Information Administration, "Net Generation by Energy Source: Total (All Sectors), 2006–December 2016," *Electric Power Monthly*, February 24, 2017.

CHAPTER 2: IF YOU ARE UNCERTAIN, CALL A QUANTUM MECHANIC FOR A FIX

1. W. Z. Heisenberg, "Über den Anschaulichen Inhalt der Quantentheoretischen Kinematik und Mechanik," *Zeitschrift für Physik* 43 (1927): 172–98.

2. Erwin Schrödinger, "An Undulatory Theory of the Mechanics of Atoms and Molecules," *Physical Review* 28, no. 6 (1926): 1049–70.

3. Erwin Schrödinger, "Die gegenwärtige Situation in der Quantenmechanik," *Naturwissenschaften* 23, no. 48 (November 1935): 807–12.

4. *Coherence*, directed by James Ward Byrkit, Oscilloscope Laboratories, 2013.

5. Hugh Everett III, "Relative State Formulation of Quantum Mechanics," *Reviews of Modern Physics* 29 (1957): 454–62.

6. Tom Siegfried, "Einstein Was Wrong about Spooky Quantum Entanglement," *Science News*, February 19, 2014, https://www.sciencenews.org/blog/context/einstein-was-wrong-about-spooky-quantum-entanglement (accessed 6/30/2017).

7. René Barjavel, *Le Voyageur Imprudent* (Éditions Denoël, 1944).

8. Zeeya Merali, "Solving Biology's Mysteries Using Quantum Mechanics," *Discover*, December 29, 2014, http://discovermagazine.com/2014/dec/17-this-quantum-life (accessed June 30, 2017).

9. Richard Hildner et al., "Quantum Coherent Energy Transfer over Varying Pathways in Single Light-Harvesting Complexes," *Science* 340, no. 6139 (2013): 1448–51.

FIRST INTERLUDE: A TOUCH OF ATOMIC THEORY

1. CERN, "The Early Universe," https://home.cern/about/physics/early-universe (accessed June 22, 2017).

2. P. A. M. Dirac, "The Quantum Theory of the Electron," *Proceedings of the Royal Society of London, Series A, Containing Papers of a Mathematical and Physical Character* 117, no. 778 (February 1, 1928): pp. 610–24.

3. Carl D. Anderson, "The Apparent Existence of Easily Deflectable Positives," *Science* 76, no. 1967 (September 9, 1932): pp. 238–39.

4. Richard Van Noorden, "Antimatter Cancer Treatment," *Chemistry World* 3, November 2006, https://www.chemistryworld.com/news/antimatter-cancer-treatment/3000368.article (accessed April 29, 2017).

5. Matt Strassler, "The Strengths of the Known Forces," Of Particular Significance, May 31, 2013, https://profmattstrassler.com/articles-and-posts/particle-physics-basics/the-known-forces-of-nature/the-strength-of-the-known-forces (accessed June 18, 2017).

6. Jim Lucas, "What is the Weak Force?" *Live Science*, December 24, 2014, https://www.livescience.com/49254-weak-force.html (accessed June 18, 2017).

CHAPTER 3: STRUMMING OUR WAY INTO EXISTENCE

1. *Doctor Who*, "Blink," season 3, episode 10, first broadcast June 9, 2007, directed by Hettie Macdonald and written by Steven Moffat.

2. Barton Zwiebach, *A First Course in String Theory* (Cambridge: Cambridge University Press, 2009).

3. Liu Cixin, *The Three-body Problem* (New York: Tom Doherty Associates, 2014).

4. China Miéville, *The City & the City* (New York: Macmillan, 2009).

5. Jeffrey A. Carver, *Sunborn* (New York: Macmillan, 2010).

6. A. Barrau et al., "Probing Loop Quantum Gravity with Evaporating Black Holes," *Physical Review Letters* 107, no. 251301 (December 16, 2011).

7. *Stanford Encyclopedia of Philosophy*, s.v. "Zeno's Paradoxes," by Nick Huggett, https://plato.stanford.edu/entries/paradox-zeno/ (accessed June 18, 2017).

CHAPTER 4: OUR UNIVERSE (AS OPPOSED TO THOSE OTHERS)

1. "How Fast Is the Universe Expanding?" NASA, https://wmap.gsfc.nasa.gov/universe/uni_expansion.html (accessed June 16, 2017).
2. *Star Wars: A New Hope*, directed by George Lucas, 20th Century Fox, 1977.
3. I. Ribas et al., "First Determination of the Distance and Fundamental Properties of an Eclipsing Binary in the Andromeda Galaxy," *Astrophysical Journal Letters* 635 (2005): L37–L40.
4. *Encyclopedia Britannica Online*, s.v. "Henrietta Swan Leavitt, American Astronomer," https://www.britannica.com/biography/Henrietta-Swan-Leavitt (accessed June 16, 2017).
5. *Third Programme*, first broadcast March 28, 1949, by BBC radio.
6. Liza Gross, "Edwin Hubble: The Great Synthesizer Revealing the Breadth and Birth of the Universe," Exploratorium, http://www.exploratorium.edu/origins/hubble/people/edwin.html (accessed June 22, 2017).
7. "Tests of Big Bang: The CMB," NASA, https://map.gsfc.nasa.gov/universe/bb_tests_cmb.html (accessed June 16, 2017).
8. C. Patrignani et al., "Big-Bang nucleosynthesis," *Chinese Physics C* 40 (2016), http://pdg.lbl.gov/2016/reviews/rpp2016-rev-bbang-nucleosynthesis.pdf (accessed June 16, 2017).
9. Lucy Hawking and Stephen Hawking, *George and the Big Bang* (New York: Doubleday Children's Books, 2011).
10. George Musser and J. R. Minkel (2002-02-11), "A Recycled Universe: Crashing Branes and Cosmic Acceleration May Power an Infinite Cycle in Which Our Universe Is but a Phase," *Scientific American* (February 2, 2002).
11. Alan H. Guth, "Inflationary Universe: A Possible Solution to the Horizon and Flatness Problems," *Physical Review D* 23, no. 347 (January 15, 1981).
12. Donald Goldsmith, "The Fingerprint of Creation," *Discover Magazine*, October 1, 1992, http://discovermagazine.com/1992/oct/thefingerprintof136 (accessed June 16, 2017).
13. Richard B. Larson and Voker Bromm, "The First Stars in the Universe," *Scientific American*, January 19, 2009, https://www.scientificamerican.com/article/the-first-stars-in-the-un/ (accessed June 16, 2017).
14. "Gravitational Waves Detected 100 Years after Einstein's Prediction," LIGO, February 11, 2016, https://www.ligo.caltech.edu/news/ligo20160211 (accessed June 16, 2017).
15. Christopher Crockett, "Cosmic Census of Galaxies Updated to 2 Trillion," *Science News*, October 12, 2016, https://www.sciencenews.org/article/cosmic-census-galaxies-updated-2-trillion (accessed June 16, 2017).
16. *Wikipedia*, s.v. "Solar Mass," https://en.wikipedia.org/wiki/Solar_mass (accessed June 16, 2017).

17. Robert Krulwich, "Which Is Greater, the Number of Sand Grains on Earth or Stars in the Sky?" National Public Radio (NPR), September 17, 2012, http://www.npr.org/sections/krulwich/2012/09/17/161096233/which-is-greater-the-number-of-sand-grains-on-earth-or-stars-in-the-sky (accessed April 27, 2017).

18. *Wikipedia*, s.v. "Milky Way," https://en.wikipedia.org/wiki/Milky_Way (accessed June 16, 2017).

19. A. M. Ghez et al., "The First Measurement of Spectral Lines in a Short-Period Star Bound to the Galaxy's Central Black Hole: A Paradox of Youth," *Astrophysical Journal* 586, no.2 (March 12, 2003): L127–L131.

20. "Where is the Ice on Ceres? New NASA Dawn Findings," NASA, December 15, 2016, https://www.jpl.nasa.gov/news/news.php?feature=6703 (accessed June 16, 2017).

21. "Sun: In Depth," NASA, https://solarsystem.nasa.gov/planets/sun/indepth (accessed June 16, 2017).

22. Fraser Cain, "How Long Does Sunlight Take to Reach Earth?" *Universe Today*, October 16, 2016, https://www.universetoday.com/15021/how-long-does-it-take-sunlight-to-reach-the-earth/ (accessed June 16, 2017).

23. "Kuiper Belt: In Depth" NASA, https://solarsystem.nasa.gov/planets/kbos/indepth (accessed June 16, 2017).

24. F. Nimmo et al., "Reorientation of Sputnik Planitia Implies a Subsurface Ocean on Pluto," *Nature* 540 (December 1, 2016): 94–96.

25. Harold F. Levison and Luke Donnes, "Comet Populations and Cometary Dynamics," in ed. Lucy Ann Adams McFadden et al. *Encyclopedia of the Solar System* (Boston: Academic Press, 2007), pp. 575–88.

26. Christopher Crockett, "The sun Isn't the Only Light Source behind That Summer Tan," *Science News*, September 20, 2016, https://www.sciencenews.org/article/sun-isn%E2%80%99t-only-light-source-behind-summer-tan (accessed June 16, 2017).

CHAPTER 5: PARALLEL WORLDS

1. *Star Trek*, "Mirror, Mirror," first broadcast October 6, 1967, on NBC, directed by Marc Daniels.

2. Charles Stross, *The Family Trade* (New York: Tor Books, 2004).

3. Peter Woit, *Not Even Wrong: The Failure of String Theory and the Search for Unity in Physical Law* (New York: Basic Books, 2006).

4. David Gerrold, *The Man Who Folded Himself* (Orbit Books, 2014).

5. David Brin, *The Practice Effect* (New York: Bantam Books, 1984).

CHAPTER 6: POWERING UP OUR CIVILIZATIONS

1. Nickolai Kardashev, "Transmission of Information by Extraterrestrial Civilizations," *Soviet Astronomy* 8 (1964): 217.
2. Freeman Dyson, "Search for Artificial Stellar Sources of Infrared Radiation," *Science* 131, no. 3414 (1960): pp. 1667–68.
3. Olaf Stapledon, *Star Maker* (London: Methuen, 1937).
4. Marc Kaufman, "The Ever More Puzzling and Intriguing, 'Tabby's Star,'" *Many Worlds*, August 8, 2016, http://www.manyworlds.space/index.php/2016/08/08/the-ever-more-puzzling-and-intriguing-tabbys-star/ (accessed June 17, 2017).
5. *David Darling Encyclopedia*, s.v. "The Milky Way Galaxy," http://www.davidarling.info/encyclopedia/G/Galaxy.html (accessed June 17, 2017).
6. "The Virgo Supercluster," *Futursim*, https://futurism.com/the-virgo-supercluster-2/ (accessed June 17, 2017).
7. Jose Luis Cordeiro, "The 'Energularity,'" Lifeboat Foundation, https://lifeboat.com/ex/the.energularity (accessed June 17, 2017).
8. M. E. Peskin and D. V. Schroeder, *An Introduction to Quantum Field Theory* (Westview Press, 1995).
9. Arnold Neumaier, "The Physics of Virtual Particles," *Physics Forums Insights*, March 28, 2016, https://www.physicsforums.com/insights/physics-virtual-particles/ (accessed June 17, 2017).
10. D. W. Sciama, "The Physical Significance of the Vacuum State of a Quantum Field," in Simon Saunders and Harvey R. Brown, eds., *The Philosophy of Vacuum* (Oxford: Oxford University Press, 1991).
11. H. B. G. Casimir, "On the Attraction between Two Perfectly Conducting Plates," *Proceedings of the Royal Netherlands Academy of Arts and Sciences* 51 (1948): 793–95.
12. S. K. Lamoreaux, "Demonstration of the Casimir Force in the 0.6 to 6 µm Range," *Physical Review Letters* 78, no. 1 (January 6, 1997): 1–4.
13. John Barrow, *Impossibility: Limits of Science and the Science of Limits* (Oxford: Oxford University Press, 1998).
14. John Barrow, *Impossibility: Limits of Science and the Science of Limits* (Oxford: Oxford University Press, 1998), p. 133.
15. Megan C. Guilford et al., "A New Long Term Assessment of Energy Return on Investment (EROI) for US Oil and Gas Discovery and Production," *Sustainability* 211, no. 3(10) (October 14, 2011): 1866–87.
16. Fraunhofer ISE, "New World Record for Solar Cell Efficiency at 46% French-German Cooperation Confirms Competitive Advantage of European Photovoltaic Industry," December 1, 2014.
17. Ben Zientara, "How Much Electricity Does a Solar Panel Produce?" *Solar*

Power Rocks, https://solarpowerrocks.com/solar-basics/how-much-electricity-does-a-solar-panel-produce/ (accessed 6/17/2017).

CHAPTER 7: BLACK HOLES SUCK

1. A. M. Ghez et al., "The First Measurement of Spectral Lines in a Short-Period Star Bound to the Galaxy's Central Black Hole: A Paradox of Youth," *Astrophysical Journal* 586, no. 2 (March 12, 2003): L127–L131.
2. Jean Tate, "Chandrasekhar Limit," *Universe Today*, December 24, 2015, https://www.universetoday.com/40852/chandrasekhar-limit/ (accessed June 19, 2017).
3. Fraser Cain, "Schwarzschild Radius," *Universe Today*, April 26, 2016, https://www.universetoday.com/39861/schwarzschild-radius/ (accessed June 19, 2017).
4. "Jupiter Fact Sheet," NASA, https://nssdc.gsfc.nasa.gov/planetary/factsheet/jupiterfact.html (accessed June 19, 2017).
5. Sean M. Carroll, *Spacetime and Geometry* (Boston: Addison-Wesley, 2004).
6. S. W. Hawking, "Black Hole Explosions," *Nature* 248 (1974): 30–31.
7. Adam Brown, "Can We Mine A Black Hole?" *Scientific American*, February 2015.
8. Ron Cowen, "The Quantum Source of Space-Time," *Nature*, November 16, 2015, http://www.nature.com/news/the-quantum-source-of-space-time-1.18797 (accessed on June 23, 2017).
9. Tony Phillips, "In Search of Gravitomagnetism," NASA, https://www.nasa.gov/vision/universe/solarsystem/19apr_gravitomagnetism.html (accessed June 23, 2017).
10. Andrew Grant, "General Relativity Caught in Action around Black Hole," *Science News*, December 17, 2015.
11. "The Most Beautiful Theory," *Economist*, November 28, 2015, http://www.economist.com/news/science-and-technology/21679172-century-ago-albert-einstein-changed-way-humans-saw-universe-his-work (accessed June 19, 2017).
12. "Behemoth Black Hole Found in an Unlikely Place," NASA, April 6, 2016, https://www.nasa.gov/feature/goddard/2016/behemoth-black-hole-found-in-an-unlikely-place (accessed June 19, 2017).

CHAPTER 8: ORIGIN AND EVOLUTION OF LIFE ON EARTH

1. Thomas Robert Malthus, *An Essay on the Principle of Population* (London: Joseph Johnson, 1798).

2. Kurt Vonnegut, *Galápagos* (New York: Dell Publishing, 1985).

3. Brendan Epstein et al., "Rapid Evolutionary Response to a Transmissible Cancer in Tasmanian Devils," *Nature Communications* 7, August 30, 2016.

4. "Age of the Earth," US Geological Survey, July 9, 2007, https://pubs.usgs.gov/gip/geotime/age.html (accessed June 21, 2017).

5. National Geographic News, "What Was 'Lucy'? Fast Facts on an Early Human Ancestor," *National Geographic*, September 20, 2006, http://news.nationalgeographic.com/news/2006/09/060920-lucy.html.

6. *10,000 BC*, directed by Roland Emmerich (Warner Brothers Pictures, 2008).

7. "'Pompeii-Like' Excavations Tell Us More about Toba Super-Eruption," *ScienceDaily*, March 3, 2010, https://www.sciencedaily.com/releases/2010/02/100227170841.htm (accessed June 21, 2017).

8. Charles Q. Choi, "DNA from Mysterious 'Denisovans' Helped Modern Humans Survive," *Live Science*, March 17, 2016, https://www.livescience.com/54084-denisovan-dna-helped-modern-humans-survive.html (accessed June 19, 2017).

9. Blake Crouch, *Pines* (Las Vegas: Thomas & Mercer, 2012).

10. Leslie Mullen, "Defining Life: Q&A with Scientist Gerald Joyce," *Space.com*, August 1, 2013, https://www.space.com/22210-life-definition-gerald-joyce-interview.html (accessed on June 23, 2017).

11. Alberto Patiño Douce, *Thermodynamics of the Earth and Planets* (Cambridge: Cambridge University Press, 2011), p. 111.

12. Bruce Alberts et al., *Molecular Biology of the Cell, 4th Edition* (New York: Garland Science, 2002).

13. Alison Abbott, "Scientists Bust Myth That Our Bodies Have More Bacteria than Human Cells," *Nature*, January 8, 2016, http://www.nature.com/news/scientists-bust-myth-that-our-bodies-have-more-bacteria-than-human-cells-1.19136 (accessed April 29, 2017).

14. "The Cambrian Explosion," PBS.org, https://www.pbs.org/wgbh/evolution/library/03/4/l_034_02.html (accessed June 21, 2017).

15. Leander Stewart et al., "Differentiating between Monozygotic Twins through DNA Methylation-Specific High-Resolution Melt Curve Analysis," *Analytical Biochemistry* 476, no. 1 (May 2015): 36–39.

16. *Star Trek: The Next Generation*, first broadcast April 26, 1993, directed by Jonathan Frakes and written by Joe Menosky.

CHAPTER 9: BADASS BIOLOGY

1. Heidi Ledford, "CRISPR: Gene Editing Is Just the Beginning," *Nature*, March 7, 2016, http://www.nature.com/news/crispr-gene-editing-is-just-the-beginning-1.19510 (accessed June 20, 2017).

2. Nancy Kress, *Beggars in Spain* (New York: William Morrow and Company, 1991).

3. Kristen Fortney et al., "Genome-Wide Scan Informed by Age-Related Disease Identifies Loci for Exceptional Human Longevity," *PLOS Genetics* (December 17, 2015).

4. Linda Marsa, "What It Takes to Reach 100," *Discover*, September 1, 2016, http://discovermagazine.com/2016/oct/what-it-takes-to-reach-100 (accessed June 21, 2017).

5. Nicola Davis, "400-Year-Old Greenland Shark Is Oldest Vertebrate Animal," *Guardian*, August 12, 2016, https://www.theguardian.com/environment/2016/aug/11/400-year-old-greenland-shark-is-the-oldest-vertebrate-animal (accessed June 19, 2017).

6. Jerry W. Shay and Woodring E. Wright, "Hayflick, His Limit, and Cellular Ageing," *Nature* (October 2000): 72–76.

7. Chanhee Kang et al., "The DNA Damage Response Induces Inflammation and Senescence by Inhibiting Autophagy of GATA4," *Science* (September 2015).

8. Madeline A. Lancaster et al., "Cerebral Organoids Model Human Brain Development and Microcephaly," *Nature* 501 (September 19, 2013): 373–79.

9. *Moon*, directed by Duncan Jones (Sony Pictures Classics, 2009).

10. *Oblivion*, directed by Joseph Kosinski (Universal Pictures, 2013).

11. Joseph Castro, "Zombie Fungus Enslaves Only Its Favorite Ant Brains," *Live Science*, September 9, 2014, https://www.livescience.com/47751-zombie-fungus-picky-about-ant-brains.html (accessed June 19, 2017).

12. Juliana Agudelo et al., "Ages at a Crime Scene: Simultaneous Estimation of the Time Since Deposition and Age of Its Originator," *Analytical Chemistry* 88, no. 12 (2016): 6479–6484.

13. Youna Hu et al., "GWAS of 89,283 Individuals Identifies Genetic Variants Associated with Self-Reporting of Being a Morning Person," *Nature Communications* February 2, 2016.

CHAPTER 10: WELCOME TO TECH U

1. Amy Ellis Nutt, "In a Medical First, Brain Implant Allows Paralyzed Man to Feel Again," *Washington Post*, October 13, 2016, https://www.washingtonpost.com/news/to-your-health/wp/2016/10/13/in-a-medical-first-brain-implant-allows-paralyzed-man-to-feel-again/?utm_term=.f39a3cf2e04b (accessed April 29, 2017).

2. Lara Lewington, "Cybathlon: Battle of the Bionic Athletes," BBC News, October 10, 2016, http://www.bbc.com/news/technology-37605984 (accessed June 19, 2017).

3. Dheeraj S. Roy et al., "Memory Retrieval by Activating Engram Cells in Mouse Models of Early Alzheimer's Disease," *Nature* 531, March 24, 2016.

4. Arthur C. Clarke, *Profiles of the Future: An Inquiry into the Limits of the Possible* (New York: Popular Library, 1973).

5. P. Kothamasu et al., "Nanocapsules: The Weapons for Novel Drug Delivery Systems," *BioImpacts* 2, no. 2 (2012): 71–81.

6. Víctor García-López et al., "Unimolecular Submersible Nanomachines. Synthesis, Actuation, and Monitoring," *Nano Letters* 15, no. 12 (2015): 8229–8239.

7. *Doctor Who*, "The Empty Child," season 1, episode 9, first broadcast May 21, 2005, directed by James Hawkes and written by Steven Moffat; *Doctor Who*, "The Doctor Dances," season 1, episode 10, first broadcast May 28, 2005, directed by James Hawkes and written by Steven Moffat.

8. Charles Stross, *Glasshouse* (New York: Ace, 2006).

9. Dan Simmons, *Ilium* (New York: HarperTorch, 2005); Dan Simmons, *Olympos* (New York: Harper Voyager, 2006).

10. John Scalzi, *Lock In: A Novel of the Near Future* (New York: Tor Books, 2014).

11. *Star Trek*, "Where No Man Has Gone Before," first broadcast September 22, 1966 on NBC, directed by James Goldstone and written by Samuel A. Peeples; *Star Trek*, "Charlie X," first broadcast September 15, 1966, on NBC, directed by Lawrence Dobkin and written by D. C. Fontana; *Star Trek*, "The Squire of Gothos," first broadcast January 12, 1967, on NBC, directed by Don McDougall and written by Paul Schneider.

12. Adi Robertson, "The Classics: 'Burning Chrome,'" *Verge*, November 3, 2012, https://www.theverge.com/2012/11/3/3594618/the-classics-burning-chrome (accessed June 19, 2017).

13. *Guillermo Fuertes et al.*, "Intelligent Packaging Systems: Sensors and Nanosensors to Monitor Food Quality and Safety," *Journal of Sensors* (2016), article ID 4046061.

CHAPTER 11: MAN AND NATURE

1. "Carbon & Tree Facts," Abor Environmental Alliance, http://www.arborenvironmentalalliance.com/carbon-tree-facts.asp (accessed on June 23, 2017).

2. National Oceanic and Atmospheric Administration, "Carbon Dioxide Levels Rose at Record Pace for 2nd Straight Year," March 10, 2017, http://www.noaa.gov/news/carbon-dioxide-levels-rose-at-record-pace-for-2nd-straight-year (accessed June 19, 2017).

3. "World of Change: Global Temperatures," NASA, https://earthobservatory.nasa.gov/Features/WorldOfChange/decadaltemp.php (accessed June 19, 2017).

4. National Centers for Environmental Information, "Global Climate Report—January 2017," https://www.ncdc.noaa.gov/sotc/global/201701 (accessed June 19, 2017).

5. Michael Slezak, "Revealed: First Mammal Species Wiped Out by Human-Induced Climate Change," *Guardian*, June 13, 2016, https://www.theguardian.com/environment/2016/jun/14/first-case-emerges-of-mammal-species-wiped-out-by-human-induced-climate-change (accessed June 19, 2017).

6. CNN Wire Staff, "Report 75% of Coral Reefs Threatened," CNN.com, March 23, 2011, http://www.cnn.com/2011/WORLD/asiapcf/02/25/world.coral.reefs/index.html (accessed June 19, 2017).

7. Amanda Mascarelli, "Climate-Change Adaptation: Designer Reefs," *Nature*, April 23, 2014, http://www.nature.com/news/climate-change-adaptation-designer-reefs-1.15073 (accessed June 19, 2017).

8. Hugh Hunt, "A Radical Proposal on Climate Change: Block out the Sun," CNN.com, June 30, 2016, http://www.cnn.com/2015/11/19/world/blocking-the-sun/index.html (accessed June 18, 2017).

9. Gaia Vince, "Sucking CO_2 From the Skies," *BBC Future*, October 4, 2012, http://www.bbc.com/future/story/20121004-fake-trees-to-clean-the-skies (accessed June 21, 2017).

10. Emily Matchar, "Will Buildings of the Future Be Cloaked In Algae?" *Smithsonian.com*, May 26, 2015, http://www.smithsonianmag.com/innovation/will-buildings-future-be-cloaked-algae-180955396/ (accessed June 21, 2017).

11. . Wl Al Sadat, "The O_2-Assisted Al/CO_2 Electrochemical Cell: A System for CO_2 Capture/Conversion and Electric Power Generation," *Science Advances* 2, no. 7, July 20, 2016.

12. Juerg M. Matter et al., "Rapid Carbon Mineralization for Permanent Disposal of Anthropogenic Carbon Dioxide Emissions," *Science* 352, no. 6291 (June 10, 2016): 1312–14.

13. David Rotman, "A Cheap and Easy Plan to Stop Global Warming," *MIT Technology Review*, February 8, 2013, https://www.technologyreview.com/s/511016/a-cheap-and-easy-plan-to-stop-global-warming/ (accessed June 21, 2017).

14. Jessica Salter, "Wrapping Greenland in Reflective Blankets," *Telegraph*, February 18, 2009, http://www.telegraph.co.uk/news/earth/environment/climatechange/4689667/Wrapping-Greenland-in-reflective-blankets.html (accessed June 21, 2017).

15. Bill Christensen, "Space-Based Sun-Shade Concept a Bright Idea," *Space.com*, November 11, 2006, https://www.space.com/3100-space-based-sun-shade-concept-bright-idea.html (accessed June 21, 2017).

16. Tobias Buckell, *Arctic Rising* (New York: St. Martins Press-3pl, 2012).

17. Paolo Bacigalupi, *The Water Knife* (New York: Knopf Doubleday, 2016).

18. Wesley Chu, *The Lives of Tao* (Nottingham, UK: Angry Robot, 2013).

19. Liu Cixin, *The Three-body Problem* (New York: Tom Doherty Associates, 2014).

20. Michel Faber, *The Book of Strange New Things: A Novel* (New York: Hogarth, 2014).

21. Holger Schmithüsen et al., "How Increasing CO_2 Leads to an Increased Negative Greenhouse Effect in Antarctica," *Geophysical Research Letters* 42, no. 23 (December 2015): 10,422–28.

22. "Ozone Destruction," Ozone Hole, http://www.theozonehole.com/ozonedestruction.htm (accessed June 19, 2017).

23. National Oceanic and Atmospheric Administration, "Scientific Assess-

ment of Ozone Depletion: 2010," https://www.esrl.noaa.gov/csd/assessments/ozone/2010/executivesummary/ (accessed June 19, 2017).

CHAPTER 12: TIME TO MOVE (PLAN B)

1. Jack Williamson, "Collision Orbit," *Astounding Science Fiction*, July 6 1942.
2. Nola Taylor Redd, "What Is Solar Wind?" *Space.com*, August 1, 2013, https://www.space.com/22215-solar-wind.html.
3. *Futurama*, "Mars University," first broadcast October 3, 1999, by Fox, directed by Bret Haaland and Gregg Vanzo and written by J. Stewart Burns.
4. Nola Taylor Redd, "How Far Away Is Venus?" *Space.com*, November 16, 2012, https://www.space.com/18529-distance-to-venus.html (accessed June 19, 2017).
5. Matt Williams, "How Do We Terraform Venus?" *Universe Today*, June 21, 2016, https://www.universetoday.com/113412/how-do-we-terraform-venus/ (accessed June 21, 2017).
6. Tim Sharp, "How Far Away Is Mars?" *Space.com*, August 2, 2012, https://www.space.com/16875-how-far-away-is-mars.html (accessed June 18, 2017).
7. Fiona MacDonald, "It's Official: NASA Announces Mars' Atmosphere Was Stripped Away by Solar Winds," *Science Alert*, November 5, 2015, https://www.sciencealert.com/live-updates-nasa-is-announcing-what-happened-to-mars-atmosphere-right-now (accessed June 19, 2017).
8. Jay Bennett, "NASA Considers Magnetic Shield to Help Mars Grow Its Atmosphere: NASA Planetary Science Division Director, Jim Green, Says Launching a Magnetic Shield Could Help Warm Mars and Possibly Allow It to Become Habitable," *Popular Mechanics*, March 1, 2017, http://www.popularmechanics.com/space/moon-mars/a25493/magnetic-shield-mars-atmosphere (accessed April 29, 2017).
9. Robert M. Zubrin and Christopher P. McKay, "Technological Requirements for Terraforming Mars," 1993, http://www.users.globalnet.co.uk/~mfogg/zubrin.htm (accessed June 21, 2017).
10. Shannon Stirone, "Your Guide to the Most Habitable Exoplanets," *Astronomy Magazine*, April 7, 2017, http://www.astronomy.com/news/2017/04/exoplanet-guide (accessed June 21, 2017).
11. Ian Sample, "Exoplanet Discovery: Seven Earth-Sized Planets Found Orbiting Nearby Star," *Guardian*, February 23, 2017, https://www.theguardian.com/science/2017/feb/22/thrilling-discovery-of-seven-earth-sized-planets-discovered-orbiting-trappist-1-star (accessed June 19, 2017).
12. Nadia Drake, "Potentially Habitable Planet Found Orbiting Star Closest to Sun," *National Geographic*, August 24, 2016, http://news.nationalgeographic.com/2016/08/earth-mass-planet-proxima-centauri-habitable-space-science/ (accessed June 19, 2017).
13. Nicola Davis, "Apollo Deep Space Astronauts Five Times More Likely to

Die from Heart Disease," *Guardian*, July 28, 2016, https://www.theguardian.com/science/2016/jul/28/apollo-deep-space-astronauts-five-times-more-likely-to-die-from-heart-disease (accessed June 19, 2017).

14. Takuma Hashimoto et al., "Extremotolerant Tardigrade Genome and Improved Radiotolerance of Human Cultured Cells by Tardigrade-Unique Protein," *Nature Communications* 7 (2016).

CHAPTER 13: INTELLIGENCE COMES IN ORGANIC AND ARTIFICIAL FLAVORS

1. F-C Yeh, "Quantifying Differences and Similarities in Whole-Brain White Matter Architecture Using Local Connectome Fingerprints," *PLoS Computational Biology* 12, no. 11, November 15, 2016, http://dx.doi.org/10.1371/journal.pcbi.1005203 (accessed April 29, 2017).

2. Isaac Asimov, *The Foundation Trilogy* (New York: Alfred A. Knopf, 2010).

3. Elizabeth Howell, "Henrietta Swan Leavitt: Discovered How to Measure Stellar Distance," *Space.com*, November 11, 2016, https://www.space.com/34708-henrietta-swan-leavitt-biography.html (accessed June 22, 2017).

4. Sara Chodosh, "The Incredible Evolution of Supercomputers' Powers, From 1946 To Today," *Popular Science*, April 22, 2017, http://www.popsci.com/supercomputers-then-and-now (accessed June 22, 2017).

5. Adrian Cho, "'Huge Leap Forward': Computer That Mimics Human Brain Beats Professional at Game of Go," *Science*, January 27, 2016, http://www.sciencemag.org/news/2016/01/huge-leap-forward-computer-mimics-human-brain-beats-professional-game-go (accessed June 22, 2017).

6. Bruce Weber, "Swift and Slashing, Computer Topples Kasparov," *New York Times*, May 12, 1997, http://www.nytimes.com/1997/05/12/nyregion/swift-and-slashing-computer-topples-kasparov.html (accessed June 19, 2017).

7. Sean O'Neill, "Forget Turing—I Want to Test Computer Creativity," *New Scientist*, December 10, 2014, https://www.newscientist.com/article/mg22429992-900-forget-turing-i-want-to-test-computer-creativity/ (accessed June 27, 2017).

8. Vernor Vinge, "The Coming Technological Singularity: How to Survive in the Post-Human Era," in *Vision-21: Interdisciplinary Science and Engineering in the Era of Cyberspace*, ed. G.A. Landis, NASA Publication CP-10129 (Washington, DC: National Aeronautics and Space Administration, 1993), pp. 11–22.

9. Andy Greenberg, "Now Anyone Can Deploy Google's Troll-Fighting AI," *Wired*, February 23, 2017, https://www.wired.com/2017/02/googles-troll-fighting-ai-now-belongs-world/ (accessed June 19, 2017).

10. Julian Jaynes, *The Origin of Consciousness in the Breakdown of the Bicameral Mind* (Boston: Houghton Mifflin, 1976).

11. Lee Bell, "What Is Moore's Law? *Wired* Explains the Theory That Defined the Tech Industry," *Wired*, August 28, 2016, http://www.wired.co.uk/article/wired-explains-moores-law (accessed June 19, 2017).

12. Tomoki W. Suzuki, Jun Kunimatsu, and Masaki Tanaka, "Correlation between Pupil Size and Subjective Passage of Time in Non-Human Primates," *Journal of Neuroscience* 2, no. 36 (November 2016): 11331–37.

13. Kim Zetter, "An Unprecedented Look at Stuxnet, the World's First Digital Weapon," *Wired*, November 3, 2014, https://www.wired.com/2014/11/countdown-to-zero-day-stuxnet/ (accessed June 19, 2017).

CHAPTER 14: THE RISE OF THE ROBOTS

1. "Jan. 25, 1921: The Robot Cometh," *Wired*, January 25, 2007, https://www.wired.com/2007/01/jan-25-1921-the-robot-cometh/ (accessed June 19, 2017).

2. Isaac Asimov, "Liar," *Astounding Science Fiction*, 1941; Alan Brown, "The Man Who Coined the Term 'Robotics,'" *From the Editors Desk* (blog), April 18, 2012, https://memagazineblog.org/2012/04/18/the-man-who-coined-the-term-robotics/ (accessed 7/1/2017).

3. Daven Hiskey, "The First Robot Was a Steam-Powered Pigeon," *Mental Floss*, http://mentalfloss.com/article/13083/first-robot-created-400-bce-was-steam-powered-pigeon (accessed June 19, 2017).

4. Ibn al-Razzaz al–Jazari, *The Book of Knowledge of Ingenious Mechanical Devices: Kitáb fí ma'rifat al-hiyal al-handasiyya*, trans. Donald R. Hill (Springer Science & Business Media, 1973).

5. Nicholas Jackson, "Elektro the Moto-Man, One of the World's First Celebrity Robots," *Atlantic*, February 21, 2011, https://www.theatlantic.com/technology/archive/2011/02/elektro-the-moto-man-one-of-the-worlds-first-celebrity-robots/71505/ (accessed June 19, 2017).

6. Hank Campbell, "Early 20th Century Robots: Sparko, the Robotic Scottish Terrier," *Science 2.0*, March 31, 2011, http://www.science20.com/science_20/early_20th_century_robots_spar...-77664 (accessed June 22, 2017).

7. Guinness World Records, "First Human to be Killed by a Robot," http://www.guinnessworldrecords.com/world-records/first-human-to-be-killed-by-a-robot (accessed June 19, 2017).

8. Kris Osborn, "Pentagon Plans for Cuts to Drone Budgets," DOD Buzz, January 2, 2014, https://www.dodbuzz.com/2014/01/02/pentagon-plans-for-cuts-to-drone-budgets/ (accessed June 22, 2017).

9. Isaac Asimov, "Runaround," *Astounding Science Fiction*, March 1942.

10. Isaac Asimov, *Foundation and Earth* (New York: Doubleday, 1986).

11. "Robot Rules, OK? An Examination of Asimov's 'Laws of Robotics' Fiction," Computer (two parts: 26, no. 12 [December 1993] and 27, no. 1 [January 1994]).

12. Gordon Briggs and Matthias Scheutz, "Sorry, I Can't Do That: Developing Mechanisms to Appropriately Reject Directives in Human-Robot Interactions," (paper presented at the AAAI Fall Symposium Series, Human Robot Interaction Laboratory, Tufts University, Medford, Massachusetts, 2015).

13. Fred Hapgood, "Chaotic Robotics," *Wired* 2, no. 9, September 1994, https://www.wired.com/1994/09/tilden/ (accessed April, 29, 2017).

14. Cecilia Kang, "Cars Talking to One Another? They Could Under Proposed Safety Rules," *New York Times*, December 13, 2016, https://www.nytimes.com/2016/12/13/technology/cars-talking-to-one-another-they-could-under-proposed-safety-rules.html?_r=0 (accessed June 19, 2017).

15. Alex Davies, "Here's What It's Like to Ride in Uber's Self-Driving Car," *Wired*, September 16, 2016, https://www.wired.com/2016/09/heres-like-ride-ubers-self-driving-car/ (accessed June 19, 2017).

16. Paul A. Eisenstein, "Now You Can Ride in a Google Self-Driving Car," NBC News, April 25, 2017, http://www.nbcnews.com/tech/tech-news/now-you-can-ride-google-self-driving-car-n750646 (accessed June 19, 2017).

17. Knvul Sheikh, "New Robot Helps Babies with Cerebral Palsy Learn to Crawl," *Scientific American*, October 1, 2016.

18. "ECCEROBOT," Technische Universität München, http://www6.in.tum.de/Main/ResearchEccerobot (accessed June 22, 2017).

19. SoftBank, "SoftBank Increases Its Interest in Aldebaran to 95%" https://www.ald.softbankrobotics.com/en/press/press-releases/softbank-increases-its-interest (accessed June 19, 2017).

20. Danielle Egan, "Here for Your Heart Surgery? Come Meet Dr. Snake-Bot," *Discover*, January 29, 2011, http://discovermagazine.com/2010/nov/29-ready-for-heart-surgery-meet-dr-snake-bot (accessed June 19, 2017).

21. Bridget Borgobello, "Knightscope Reveals Robotic Security Guard," *New Atlas*, December 5, 2013, http://newatlas.com/knightscope-k5-k10-robot-security-guard/30024/ (accessed June 19, 2017).

CHAPTER 15: ARE WE ALONE? EXTRATERRESTRIAL INTELLIGENCE

1. Harry Bates, "Farewell to the Master," *Astounding Science Fiction*, 1940.

2. "The Post-Detection SETI Protocol," North American Astrophysical Observatory, http://www.naapo.org/SETIprotocol.htm (accessed June 30, 2017).

3. "How Many Solar Systems Are in Our Galaxy"? NASA, https://spaceplace.nasa.gov/review/dr-marc-space/solar-systems-in-galaxy.html (accessed June 19, 2017).

4. "The Nobel Prize in Physics," Nobel Prizes and Laureates, https://www.nobelprize.org/nobel_prizes/physics/laureates/1938/ (accessed June 19, 2017).

5. "Stars and Habitable Planets," Sol, http://www.solstation.com/habitable.htm (accessed June 19, 2017).

6. Charles Q. Choi, "Double Sunsets May be Common, but Twin-Star Setups Still Mysterious," *Space.com*, January 18, 2010, https://www.space.com/7792-double-sunsets-common-twin-star-setups-mysterious.html (accessed June 22, 2017).

7. *Star Trek*, "The Devil in the Dark," first broadcast March 9, 1967, by NBC, directed by Joseph Pevney and written by Gene L. Coon.

8. Josef Allen Hynek, *The UFO Experience: A Scientific Inquiry* (Chicago: H. Regnery Company, 1972).

9. Nell Greenfieldboyce, "NASA Spots What May Be Plumes of Water on Jupiter's Moon Europa," NPR the Two-Way, September 26, 2016, http://www.npr.org/sections/thetwo-way/2016/09/26/495512651/nasa-spots-what-may-be-plumes-of-water-on-jupiters-moon-europa (accessed June 22, 2017).

10. "Post-Detection SETI."

CHAPTER 16: A REALLY LONG-DISTANCE CALL: INTERSTELLAR COMMUNICATION

1. *Encyclopedia of Science Fiction*, s.v. "Ansible."

2. "Taming Photons, Electrons Paves Way for Quantum Internet," *China Technology News*, September 20, 2016, http://www.technologynewschina.com/2016/09/taming-photons-electrons-paves-way-for.html (accessed June 19, 2017).

3. Ling Xin, "China Launches World's First Quantum Science Satellite," *IOP Physics World*, August 16, 2016, http://physicsworld.com/cws/article/news/2016/aug/16/china-launches-world-s-first-quantum-science-satellite (accessed April 29, 2017).

CHAPTER 17: *AD ASTRA PER ASPERA*: A ROUGH ROAD LEADS TO THE STARS

1. *Serenity*, directed by Joss Whedon, Universal Pictures, 2005.

2. Jacob Astor IV, *A Journey in Other Worlds: A Romance of the Future* (D Appleton, 1894).

3. Frank K. Kelly, "Star Ship Invincible," *Astounding Stories* (1935).

4. H. G. Wells, *The First Men in the Moon* (Marblehead, Massachusetts: Trajectory, 2014).

5. "Escape Velocity: Fun and Games," NASA, https://www.nasa.gov/audience/foreducators/k-4/features/F_Escape_Velocity.html (accessed June 23, 2017).

6. *New World Encyclopedia*, s.v. "Space Elevator," http://www.newworld encyclopedia.org/entry/Space_elevator (accessed June 30, 2017).

7. Arthur C. Clarke, *The Fountains of Paradise* (New York: Hartcourt Brace Jovanovich, 1979).

8. Stuart Fox, "How Do Solar Sails Work?" *Live Science*, May 17, 2010, https://www.livescience.com/32593-how-do-solar-sails-work-.html (accessed June 23, 2017).

9. "Space 'Spiderwebs' Could Propel Future Probes," *New Scientist*, April 25, 2008, https://www.newscientist.com/article/dn13776-space-spiderwebs-could-propel-future-probes/ (accessed July 2, 2017).

10. Dan Simmons, *Ilium* (New York: HarperTorch, 2005); Dan Simmons, *Olympos* (New York: Harper Voyager, 2006).

11. Gerald P. Jackson and Steven D. Howe, "Antimatter Driven Sail for Deep Space Missions," (Proceedings of the Particle Accelerator Conference, Portland, OR, May 12–16, 2003).

12. *New World Encyclopedia*, s.v. "Antimatter," http://www.newworld encyclopedia.org/entry/Antimatter (accessed June 23, 2017).

13. Natalie Wolchover, "Will Antimatter Destroy the World?" *Live Science*, June 16, 2011, https://www.livescience.com/33348-antimatter-destroy-world.html (accessed June 24, 2017).

14. Matt Williams, "What Is the Alcubierre 'Warp' Drive," *Universe Today*, Jan. 22, 2017, https://www.universetoday.com/89074/what-is-the-alcubierre-warp-drive (accessed June 24, 2017).

15. "Eugene Podkletnov's Gravity Impulse Generator," *American Antigravity*, September 21, 2012, http://www.americanantigravity.com/news/space/eugene-podkletnovs-gravity-impulse-generator.html (accessed June 23, 2017).

16. Breakthrough Initiatives, "Internet Investor and Science Philanthropist Yuri Milner & Physicist Stephen Hawking Announce Breakthrough Starshot Project to Develop 100 Million Mile per Hour Mission to the Stars within a Generation," https://breakthroughinitiatives.org/News/4 (accessed June 23, 2017).

THIRD INTERLUDE: A MATTER OF SUBSTANCE

1. Brian Greene, "How the Higgs Boson Was Found," *Smithsonian Magazine*, July 2013, http://www.smithsonianmag.com/science-nature/how-the-higgs-boson-was-found-4723520/ (accessed June 23, 2017).

2. Clara Moskowitz, "What's 96 Percent of the Universe Made Of? Astronomers Don't Know," *Space.com*, May 12, 2011, https://www.space.com/11642-dark-matter-dark-energy-4-percent-universe-panek.html (accessed June 23, 2017).

3. William J. Cromie, "Physicists Slow Speed of Light," *Harvard Gazette*, February 18, 1999, http://news.harvard.edu/gazette/story/1999/02/physicists-slow-speed-of-light/ (accessed April 29, 2017).

4. Julian Léonard et al., "Supersolid Formation in a Quantum Gas Breaking a Continuous Translational Symmetry," *Nature* 543 (March 2, 2017): 87–90.

5. Jun-Ru Li et al., "A Stripe Phase with Supersolid Properties in Spin–Orbit-Coupled Bose–Einstein Condensates," *Nature* 543 (March 2, 2017): 91–94.

CHAPTER 18: WHY ARE WE SO MATERIALISTIC?

1. Richard Gray, "No More Smashed Phones! Super-Hard Metallic Glass Is 600 Times Stronger than Steel and Will BOUNCE If It's Dropped," *Daily Mail*, April 5, 2016, http://www.dailymail.co.uk/sciencetech/article-3524128/No-smashed-phones-Super-hard-metallic-glass-500-times-stronger-steel-BOUNCE-dropped.html (accessed June 23, 2017).

2. "Digital Contact Lenses Can Transform Diabetes Care," *Medical Futurist*, http://medicalfuturist.com/googles-amazing-digital-contact-lens-can-transform-diabetes-care/ (accessed June 23, 2017).

3. Melissa Healy, "Hot? You Can Cool Down by Suiting Up in This High-Tech Fabric," *Los Angeles Times*, September 1, 2016, http://www.latimes.com/science/sciencenow/la-sci-sn-cool-shirt-20160901-snap-story.html (accessed April 29, 2017).

4. Jennifer Chu, "Beaver-Inspired Wetsuits in the Works," *MIT News*, October 5, 2016, http://news.mit.edu/2016/beaver-inspired-wetsuits-surfers-1005 (accessed April 29, 2017).

5. Jyllian Kemsley, "Names for Elements 113, 115, 117, and 118 Finalized by IUPAC," *Chemical & Engineering News*, December 5, 2016, http://cen.acs.org/articles/94/i48/Names-elements-113-115-117.html (accessed 6/24/ 2017).

6. Bob Yirka, "Linking Superatoms to Make Molecules to Use as Building Blocks for New Materials," *Phys.org*, July 27, 2016, https://phys.org/news/2016-07-linking-superatoms-molecules-blocks-materials.html (accessed June 23, 2017).

7. Sarah Zielinski, "Absolute Zero: Why Is A Negative Number Called Absolute Zero?" *Smithsonian Magazine*, January 1, 2008, http://www.smithsonianmag.com/science-nature/absolute-zero-13930448 (accessed June 24, 2017).

8. Emily Conover, "The Pressure Is on to Make Metallic Hydrogen," *Science News* 190, no. 4 (August 20, 2016): 18, https://www.sciencenews.org/article/pressure-make-metallic-hydrogen (accessed April 29, 2017).

9. Colin Barras, "Warmest Ever Superconductor Works at Antarctic Temperatures," *New Scientist*, August 17, 2015, https://www.newscientist.com/article/dn28058-warmest-ever-superconductor-works-at-antarctic-temperatures/ (accessed June 23, 2017).

10. "The Nobel Prize in Physics 2010," Nobelprize.org, https://www.nobelprize.org/nobel_prizes/physics/laureates/2010/ (accessed June 23, 2017).

11. Prachi Patel, "Silkworms Spin Super-Silk after Eating Carbon Nanotubes and Graphene," *Scientific American*, October 9, 2016, https://www.scientific

american.com/article/silkworms-spin-super-silk-after-eating-carbon-nanotubes-and-graphene/ (accessed April 29, 2017).

12. Lynda Delacey, "Q-Carbon: A New Phase of Carbon So Hard It Forms Diamonds When Melted," *New Atlas*, December 6, 2015, http://newatlas.com/q-carbon-new-phase-of-carbon/40668/ (accessed June 23, 2017).

13. Alexandra Goho, "Infrared Vision: New Material Might Enhance Plastic Solar Cells," *Science News*, January 22, 2005.

14. Stefan Lovgren, "Spray-On Solar-Power Cells Are True Breakthrough," *National Geographic News*, January 14, 2005, http://news.nationalgeographic.com/news/2005/01/0114_050114_solarplastic.html (accessed June 23, 2017).

15. Goho, "Infrared Vision."

16. Seth Borenstein, "Now You See It, Now You Don't: Time Cloak Created," *US News*, January 4, 2012, https://www.usnews.com/science/articles/2012/01/04/now-you-see-it-now-you-dont-time-cloak-created (accessed April 29, 2017).

CHAPTER 19: TECHNOLOGY (COOL TOYS)

1. Anh-Vu Do et al., "3D Printing of Scaffolds for Tissue Regeneration Applications," *Advanced Healthcare Materials* 4, no. 12 (2015): 1742–62.

2. William Herkewitz, "Incredible 3D Printer Can Make Bone, Cartilage, and Muscle. Hello, Future," *Popular Mechanics*, February 15, 2016, http://www.popularmechanics.com/science/health/a19443/3d-printer-bone-cartilidge-and-muscle/ (accessed April 29, 2017).

3. Steven Leckart, "How 3-D Printing Body Parts Will Revolutionize Medicine," *Popular Science*, August 6, 2013, http://www.popsci.com/science/article/2013-07/how-3-d-printing-body-parts-will-revolutionize-medicine (accessed June 23, 2017).

4. Marissa Fessenden, "3-D Printed Windpipe Gives Infant Breath of Life: A Flexible, Absorbable Tube Helps a Baby Boy Breathe, and Heralds a Future of Body Parts Printed on Command," *Scientific American*, May 24, 2013, https://www.scientificamerican.com/article/3-d-printed-windpipe/ (accessed April 29, 2017).

5. Alexandria Le Tellier, "Does the Bluetooth Dress Signal the Future of Fashion?" *All the Rage*, June 18, 2009, http://latimesblogs.latimes.com/alltherage/2009/06/does-the-bluetooth-dress-signal-the-future-of-fashion.html (accessed June 23, 2017).

6. Richard Gray for *MailOnline*, "The Electronic Skin Fitted with 'Disco Lights': Sticky Film Could Lead to Wearable Screens That Track Your Health and Even Show FILMS," *DailyMail*, April 15, 2016, http://www.dailymail.co.uk/sciencetech/article-3542072/The-electronic-skin-fitted-disco-lights-Sticky-film-lead-wearable-screens-track-health-FILMS.html (accessed April 29, 2017).

7. Meghan Rosen, "Tracking Health Is No Sweat with New Device: Wearable Electronic Analyzes Chemicals in Perspiration," *Science News*, January 27, 2016,

https://www.sciencenews.org/article/tracking-health-no-sweat-new-device?mode =magazine&context=543 (accessed April 29, 2017).

8. Jacob Aron, "Laser Camera Can Track Hidden Moving Objects around Corners," *New Scientist*, December 7, 2015, https://www.newscientist.com/article/ dn28628-laser-camera-can-track-hidden-moving-objects-around-corners/ (accessed April 29, 2017).

9. Emily Conover, "Wi-Fi Can Help House Distinguish between Members: Smart Homes Will Cater to Individuals' Needs," *Science News*, September 27, 2016, https://www.sciencenews.org/article/wi-fi-can-help-house-distinguish-between -members?mode=topic&context=96 (accessed April 29, 2017).

10. Dr. Elizabeth Strychalski, "Neural Engineering System Design (NESD)," DARPRA, http://www.darpa.mil/program/neural-engineering-system-design (accessed April 29, 2017).

CHAPTER 20: WHAT DOES IT TAKE TO EXIST?

1. Stephen Hawking and Leonard Mlodinow, *The Grand Design* (New York: Bantam Books, 2010).

2. *Doctor Who*, "Under the Lake," season 9, episode 3, first broadcast October 3, 2015, directed by Daniel O'Hara and written by Toby Whithouse.

3. "The First Flight Simulator (1929)," HistoryofInformation.com, http:// www.historyofinformation.com/expanded.php?id=2520 (accessed June 23, 2017).

4. "History of Virtual Reality," Virtual Reality Society, https://www.vrs.org.uk/ virtual-reality/history.html (accessed June 23, 2017).

5. Ibid.

6. R. Gonçalves et al., "Efficacy of Virtual Reality Exposure Therapy in the Treatment of PTSD: A Systematic Review," *PLoS ONE* 7, no. 12 (2012).

7. Repstein, "Perspective; Chapter 1: The Party," Expanded Theater, September 29, 2015, https://courses.ideate.cmu.edu/54-498/f2015/perspective -chapter-1-the-party-by-morris-may-and-rose-troche-2015/ (accessed June 23, 2017); Angela Watercutter, "VR Films Are Going to Be All over Sundance in 2015," *Wired*, December 4, 2014, https://www.wired.com/2014/12/oculus-rift-sundance-film -festival/ (accessed June 23, 2017).

8. Kathryn Y. Segovia and Jeremy N. Bailenson, "Virtually True: Children's Acquisition of False Memories in Virtual Reality," *Media Psychology* 12 (2009): 371–393.

9. Jerry Adler, "Erasing Painful Memories," *Scientific American*, May 2012.

10. "Holographic Universe," *Science Daily*, https://www.sciencedaily.com/ terms/holographic_principle.htm (accessed June 23, 2017).

11. *Star Trek: The Next Generation*, "Relics," first broadcast October 12, 1992, directed by Alexander Singer and written by Ronald D. Moore.

CHAPTER 21: THE END OF EVERYTHING

1. Peter Holley, "Stephen Hawking Just Gave Humanity a Due Date for Finding another Planet," *Washington Post*, November 17, 2016, https://www.washingtonpost.com/news/speaking-of-science/wp/2016/11/17/stephen-hawking-just-gave-humanity-a-due-date-for-finding-another-planet/?utm_term=.9258c8943376 (accessed June 23, 2017).

2. Martin Rees, *Our Final Hour: A Scientist's Warning—How Terror, Error, and Environmental Disaster Threaten Humankind's Future in This Century—On Earth and Beyond* (New York: Basic Books, 2003).

3. "Roundtable: A Modern Mass Extinction?" Evolution, http://www.pbs.org/wgbh/evolution/extinction/massext/statement_03.html (accessed June 23, 2017).

4. *Wikipedia*, s.v. "Extinction Event," https://en.wikipedia.org/wiki/Extinction_event (accessed June 24, 2017).

5. Gerardo Ceballos et al., "Accelerated Modern Human-Induced Species Losses: Entering the Sixth Mass Extinction," *Science Advances* 1 (June 19, 2015).

6. Adrienne Lafrance, "The Chilling Regularity of Mass Extinctions: Scientists Say New Evidence Supports a 26-Million-Year Cycle Linking Comet Showers and Global Die-Offs," *Atlantic*, November 3, 2015, https://www.theatlantic.com/science/archive/2015/11/the-next-mass-extinction/413884/ (accessed April 29, 2017).

7. Konstantin Batygin and Michael E. Brown, "Evidence for a Distant Giant Planet in the Solar System," *Astronomical Journal* 20 (2016).

8. Leslie Mullen, "Getting Wise about Nemesis," *Astrobiology*, March 11, 2010, http://www.astrobio.net/news-exclusive/getting-wise-about-nemesis/ (accessed April 29, 2017).

9. Sarah Scoles, "Target Earth: The Next Extinction from Space," *Discover*, July 28, 2016, http://discovermagazine.com/2016/sept/9-death-from-above (accessed April 29, 2017).

10. D. C. Agle et al., "Catalog of Known Near-Earth Asteroids Tops 15,000," (announcement by Jet Propulsion Laboratory, California Institute of Technology, Pasadena, CA, October 27, 2016), https://www.jpl.nasa.gov/news/news.php?feature=6664 (accessed April 29, 2017).

11. "What Killed the Dinosaurs?" Evolution, http://www.pbs.org/wgbh/evolution/extinction/dinosaurs/asteroid.html (accessed June 23, 2017).

12. Andrey Kuzmin, "Meteorite Explodes over Russia, More Than 1,000 Injured," Reuters, February 15, 2013, http://www.reuters.com/article/us-russia-meteorite-idUSBRE91E05Z20130215 (accessed June 23, 2017).

13. Deborah Byrd, "Night Sky as Milky Way and Andromeda Galaxies Merge," EarthSky, March 24, 2014, http://earthsky.org/space/video-of-earths-night-sky-between-now-and-7-billion-years (accessed June 23, 2017).

14. "The Three Laws of Thermodynamics," *Boundless.com* https://www

.boundless.com/chemistry/textbooks/boundless-chemistry-textbook/thermodynamics-17/the-laws-of-thermodynamics-123/the-three-laws-of-thermodynamics-496-3601/ (accessed June 26, 2017); Jim Lucus, "What Is Thermodynamics," Live *Science*, May 7, 2015, https://www.livescience.com/50776-thermodynamics.html (accessed June 26, 2017).

15. Hannah Devlin, "This is the Way the World Ends: Not with a Bang, but with a Big Rip," *Guardian*, July 3, 2015, https://www.theguardian.com/science/2015/jul/02/not-with-a-bang-but-with-a-big-rip-how-the-world-will-end (accessed July 1, 2017).

16. Arthur C. Clarke, *Profiles of the Future: An Inquiry into the Limits of the Possible* (New York: Popular Library, 1973).

GLOSSARY

abiogenesis: The evolution of life from nonliving substances.

absolute zero: The theoretical lowest temperature where the motion of particles is at their minimum. Superconductors love this temperature.

adaptation: The short-run process of change by which an organism becomes better suited to the environment.

albedo: The proportion of light and heat reflected by a surface.

anthropic principle (more philosophical than scientific): Our universe is hospitable to life because we are here to observe it.

artificial gravity: I'm guessing gravity not created by stellar objects. It seems to me that, human-made or not, gravity is still gravity. What's artificial about it?

artificial intelligence: A computer system able to perform tasks that normally require human intelligence.

augmented reality: Computer-generated images superimposed over a user's view of the real world.

bacteriophages: A virus that infects and multiplies within a bacterium.

beta decay: Decay kicked off by the presence of too many protons or neutrons in an atom's nucleus. One of the excess particles will transform into the other (electrons and neutrinos are players in the process).

big bang: The leading cosmological theory on how our universe began.

carbon dioxide (CO_2): A molecule that absorbs and emits solar radiation. It is a greenhouse gas that is vital to life, but too much of anything can be bad.

carbon sequestering: It's a process that removes CO_2 from the atmosphere and stores it.

Casimir effect: The attractive force between two close plates caused by vacuum pressure. Experimentally, this provides evidence of virtual particles.

Cepheid stars: A star that pulsates at regular intervals. Mix in a little math, and it helps determine stellar distances.

Chandrasekhar limit: Calculated at 1.4 times the sun. If the mass of a star is less than this, it will end up as a white dwarf at the end of its life. If the mass is greater, things really heat up to supernova level. If a star's mass is way above this limit, it is destined to become a black hole.

chemical rocket: Uses chemical fuel and an oxidizer for combustion.

cloning: Asexual reproduction.

cognition: The acquisition of knowledge through thought.

consciousness: Awareness of your surroundings.

cosmic microwave background (CMB): The universe's first light. It is the electromagnetic radiation left over from the big bang. Think of it as a baby picture.

cosmology: The science of the whole universe, including its origin and ending.

cyberspace: A computer network environment over which users can communicate.

cyborg: A being composed of both biological and mechanical body parts.

decoherence: The transition from a quantum state into a classical state; a collapse of the probability wave into a single state (outcome).

DNA (deoxyribonucleic acid): The genetic blueprint for self-replication.

Drake equation: An equation used to calculate the odds of finding an extraterrestrial intelligent communicating civilization.

element: A substance that can't be broken down into simpler substances (they are composed of quarks and electrons).

emergent properties: A system where the whole has a property none of its individual parts have.

energy and mass equivalency ($E=mc^2$): Mass is very concentrated energy.

entropy: The amount of thermal energy unavailable for mechanical work. Also a measure of disorder in a system.

event horizon: A boundary surrounding a black hole. Once past it, there is no return, even for light.

evolution: The long-run process of inheritable changes (random mutations) of a population over time.
exoplanet: A planet outside our solar system.
Fermi Paradox: The contradiction between the high probability of alien life and the lack of evidence for that life.
general relativity: Mix gravity with special relativity, and you will discover that gravity and acceleration are equivalent.
genetic engineering: Modifying an organism through gene manipulation.
geoengineering: Human manipulation of a planet's environment.
Goldilocks zone: The *just right* distance from a star for a planet to possibly have liquid water.
gravitational waves: Ripples in spacetime caused by some wicked cosmic events like black holes colliding.
gravitons: Theorized quantized gravity (a particle) that mediates the gravitational field.
gravity-acceleration equivalence: A conclusion of general relativity that states gravity comes from acceleration.
gravity as geometry: Einstein's idea that gravity is not a force but a property of spacetime's curves. Accelerate down a curve, and boom! Gravity (see previous term).
gravity tractor: A theoretical object that can deflect another object using its gravitational field.
greenhouse effect: A metaphor used to explain the correlation of global warming and solar heat trapped in the atmosphere.
Hawking radiation: Predicted radiation that leaks from a black hole thanks to quantum effects.
Hayflick limit: A limit on the number of times a normal cell divides until it says, "No more."
heat death: The ultimate fate of the universe. No thermal energy is available for any process.
Heisenberg uncertainty principle: There is fuzziness in nature that limits what can be known about the behavior of quantum particles.
holographic principle: A theory where everything we see in spacetime is really a lower-dimensional projection from a boundary.
hominid: Hominins plus apes, gorillas, chimps, and orangutans.

hominin: Humans and any early subspecies of humans.
Hubble's constant: The rate of expansion of the universe. It is about 70.8 kilometers per second per megaparsec.
intelligence: The accumulation of knowledge.
ion engine: Ionized gas for propulsion.
Kardashev scale: Ranking system of civilizations based on power usage.
laser: Focused and amplified light emissions.
Local Group: About fifty-four gravitationally bound galaxies. The Milky Way is a member.
loop quantum gravity (LQG) theory: A quantum version of spacetime. Instead of spacetime being continuous, it might be composed of tiny (Planck length) spacetime pixels.
macrophage: A large white blood cell that ingests foreign particles, including bacteria.
membrane theory: A cosmology in which our universe is camped out on a membrane (brane) of energy that floats in a higher-dimensional bulk made up of many branes. Gravity leaks through the bulk, making gravity appear weak compared to the other three universal forces.
metamaterials: Material with properties not found in natural materials.
nanites: A nanomachine operating at the nanoscale.
natural selection: The process where those members of a species that are better adapted are more likely to survive, meet a nice partner, and have a baby.
near-Earth objects (NEOs): Asteroids and comets in Earth's neighborhood (within 1.3 AU).
neurons: A nerve cell used to transmit impulses.
neutrino: A neutrally charged particle produced by beta decay.
nuclear propulsion (Orion design): What it sounds like—nuclear explosions for thrust. A more advanced version of nuclear propulsion for rockets would require controlled nuclear fission or fusion drives. This is still in the developmental stage.
parsec: Shorthand for parallax of one arcsecond, a measure of stellar distances.
Planck length: A length so small that classical ideas about physics

are no longer valid. Quantum mechanics dominates. If strings vibrate particles into existence, then they will exist at or below this length.

posthumanism: A state of being beyond human.

quantum entanglement: Occurs when pairs of particles share the same quantum state and cannot be described independently from each other. Distance between the particles doesn't matter.

quantum teleportation: The transfer of quantum information between different locations.

ramjet: Instead of sucking air, the ramjet rams into it. The air compression ignites the fuel to produce thrust. Combustion occurs at subsonic speeds.

Rare Earth Hypothesis: Suggests that earthlike planets with complex life are rare.

redshift: A stretching of a wavelength to the red part of the electromagnetic spectrum as a stellar object moves away from an observer. This is to light what the Doppler shift is to sound.

RNA (ribonucleic acid): The worker that carries out the instructions found in the DNA genetic blueprint.

robot: Not an AI.

Schrödinger's cat: An unhappy feline used in a thought experiment.

Schrödinger's wave function: A math construct that contains all the probabilities for an outcome.

Schwarzschild radius: The radius of a given mass where, if the mass could be compressed to fit within this radius, no force could stop it from continuing its collapse to a singularity.

scramjet: A supersonic ramjet. Unlike a standard ramjet, it doesn't need to slow to subsonic speeds before combustion (and, you know, thrust).

singularity: Not clearly defined in physics, but the math of general relativity describes it as an infinitely small and dense point.

solar radiation: Solar rays identified by their wavelength.

solar sail: Uses the pressure from sunlight hitting mirrors for propulsion.

solar winds: Charged particles flowing from the sun.

space elevator: A cable anchored to a planet with its terminus in space. This could be used for ferrying equipment into orbit.

spacetime: The fusion of the three spatial dimensions of space with time into a single continuum.

special relativity: Laws of motion are the same for non-accelerating frames of reference. Speed of light is the same for all reference frames. These two facts combined yield a lot of the funky stuff in the first chapter.

speciation: The formation of unique species over the course of evolution.

stem cell: An undifferentiated cell capable of differentiating into specialized cells.

string theory: A theory that one-dimensional strings vibrating in multiple dimensions interact with each other to create science fiction along with everything else.

superatom: A group of atoms that when combined have some of the properties of an element.

superconductor: It conducts electricity in a super way (with no loss of energy).

super-galaxy: A merger of gravitationally bound galaxies into a single galaxy.

supernova: Stellar explosions. Also good for determining stellar distances.

superposition: Simultaneously maintaining all the possible states (outcomes) in the probability wave function.

tachyon particle: A theoretical particle that must always travel faster than the speed of light. The particle is inconsistent with relativity only if it slowed to subluminal speeds.

technology: The application of science for practical purposes and some not so practical but fun purposes.

terraforming: Planetary engineering to make a planet more earthlike.

thermodynamics: The physics of heat, energy, and work in a system.

3-D printing: Creation of a 3-D object from a digital model. The desired object is created in a succession of layers.

time dilation: Clocks in different frames of reference (different accelerating frames) move at different speeds relative to each other. This is one of the funky traits of special relativity.

transhumanism: The human in transition using available biological or mechanical enhancements.

transit method: A technique using measurements of light to detect exoplanets.
turbojet: Sucks in air for combustion to produce thrust.
vacuum energy/zero-point energy: The universe's background energy.
virtual particles: Short-lived particles birthed from temporary violations of the conservation of energy. It is all legal thanks to the Heisenberg uncertainty principle.
virtual reality: An interactive three-dimensional environment that appears (feels) real.
volatiles: Elements with low boiling points such as nitrogen and carbon dioxide.
wetware: Technology that links the human brain to artificial systems.
wobble method: A technique using measurements of gravitational pull to detect exoplanets.
zombie: Reanimated dead, or a living being made into an automaton (fiction).

READING/MOVIE/SONG LIST

The chapter the book or movie is relevant to is listed in parentheses.

READING LIST

Edwin A. Abbott. *Flatland: A Romance of Many Dimensions.* London: Seeley, 1884. An examination of geometric dimensions and Victorian culture. (Chapter 3.)

Douglas Adams. *The Hitchhiker's Guide to the Galaxy.* London: Pan Books, 1979. Humans evolved as part of a computer program to determine the question to life, the universe, and everything. (Chapters 8 and 13.)

Isaac Asimov. *The Foundation Trilogy.* New York: Alfred A. Knopf, 2010. Making a plan and sticking to it. (Chapter 13.)

Isaac Asimov. "Liar." *Astounding Science Fiction* 27, no. 3 (May 1941). This story contains the first use of the word "robotics." (Chapter 14.)

Isaac Asimov. "Runaround." *Astounding Science Fiction* 29, no. 1 (March 1942). Tons of robot goodness including the Three Laws of Robotics. (Chapter 14.)

Jacob Astor IV. *A Journey in Other Worlds: A Romance of the Future.* New York: D. Appleton, 1894. The first appearance of the term "spaceship." (Chapter 17.)

Margaret Atwood. *The Handmaid's Tale.* Toronto: McClelland & Stewart, 1985. This is an example of the relationship between our bodies, our environment, and politics. Human effects on the environment have caused human infertility. Non-barren women are a limited resource to be managed. (Chapter 11.)

Paolo Bacigalupi. *The Water Knife*. New York: Knopf Doubleday, 2016. The American Southwest is water-depleted, and waring nations fight over what's left of the Colorado River. (Chapter 11.)

René Barjavel. *Le Voyageur imprudent*. Paris: Éditions Denoël, 1944. The introduction of the grandfather paradox. (Chapter 2.)

Harry Bates. "Farewell to the Master." *Astounding Science Fiction* 26, no. 2 (October 1940). This story is the blueprint for the movie *The Day the Earth Stood Still*. It is a story of first contact. (Chapter 15.)

Stephen Baxter. *Evolution*. London: Orion, 2002. The title says it all. (Chapter 8.)

Greg Bear. *Eon*. New York: Tor Books, 1985. This book has everything: genetic engineering, parallel universes, and most other things from various chapters.

Gregory Benford. *Timescape*. New York: Simon & Schuster, 1980. Tachyons are used for time-traveling communication to warn of ecological disaster. (Chapters 6 and 11.)

David Brin. *The Practice Effect*. New York: Bantam Books, 1984. The main character travels to an alternate universe with different laws of physics. (Chapter 5.)

Max Brooks. *World War Z: An Oral History of the Zombie War*. New York: Three Rivers, 2007. In this book it is the Solanum virus that makes human flesh irresistible to the zombie community.

Tobias Buckell. *Arctic Rising*. New York: St. Martin's Press-3pl, 2012. A novel about the international relations after the loss of ice in the Arctic Ocean. (Chapter 11.)

Karel Capek. *R.U.R. Rossum's Universal Robots* (play). Aventinum, 1920. From this story we get the word "robot." Robot is the Czech word for "worker." (Chapter 14.)

Orson Scott Card. *The Ender Quintet*. New York: Tor Science Fiction; Box edition, 2008. This includes relativistic time travel and interstellar communication. (Chapters 1 and 16.)

M. R. Carey. *The Girl with All the Gifts*. London: Orbit Books, 2014. A novel about how messing with the environment will have biological consequences: zombies. (Chapter 9.)

Jeffrey A. Carver. *Sunborn* (the fourth book of *The Chaos Chronicles*). London: Macmillan, 2010. This book contains ancient

AIs (chapter 13) living within compact dimensions (chapter 3) inside a black hole (chapter 7).

Ted Chiang. "Story of Your Life." *Starlight 2*. Edited by Patrick Nielsen Hayden. New York: Tor Books, 2002. This includes lingual determinism and alien contact. This novella is the basis for the movie *Arrival*. (Chapter 15.)

Wesley Chu. *The Lives of Tao*. Nottingham, UK: Angry Robot, 2013. Aliens use humans to increase greenhouse gasses to make the earth more suitable for their takeover. (Chapter 11.)

Liu Cixin. *The Three-Body Problem*. New York: Tor Books, 2014. It has examples of higher and lower dimensions beyond the ones we can perceive. It is also about planetary environmental problems. (Chapters 3 and 11.)

Arthur C. Clarke. *2001: A Space Odyssey*. New York: New American Library, 1968. There is contact with an extraterrestrial intelligence, a possibly psychotic artificial intelligence, and displays of artificial gravity. (Chapters 13, 17, and 19.)

Arthur C. Clarke. *The Fountains of Paradise*. San Diego: Harcourt Brace Jovanovich, 1979. Early fictional use of the space elevator. (Chapter 17.)

Arthur C. Clarke. *Imperial Earth*. London: Gollancz, 1975. Use of smartphones and internet. Of course, they had different names because this was published in 1975. (Chapter 19.)

Ernest Cline. *Ready Player One*. New York: Random House, 2011. Virtual reality gaming for a big prize and survival. (Chapter 20.)

Blake Crouch. *Pines* (Book One of the Wayward Pines Trilogy). Seattle: Thomas & Mercer, 2012. This is about evolution gone haywire. (Chapter 8.)

Paige Daniels. "The Outpost." *Brave New Girls: Tales of Girls and Gadgets*. Edited by Paige Daniels and Mary Fran. CreateSpace, 2015. A YA story jam-packed with wormholes. (Chapter 1.)

Samuel R. Delany. *Triton*. New York: Bantam Books, 1976. An interesting spin on antigravity. (Chapter 17.)

Philip K. Dick. *Do Androids Dream of Electric Sheep?* New York: Doubleday, 1968. (The basis for the movie *Blade Runner*.) The story questions what it means to be human. (Chapter 13.)

Philip K. Dick. "We Can Remember It for You Wholesale." *Magazine* of Fantasy & Science Fiction 30, no. 4 (April 1966). This story was loosely adapted into the movie *Total Recall*. It is all a question of identity. (Chapter 20.)

Michel Faber. *The Book of Strange New Things: A Novel*. London: Hogarth, 2014. Humans fleeing a dying Earth cause human-style environmental havoc on an inhabited alien world. (Chapter 11.)

David Gerrold. *The Man Who Folded Himself*. London: Orbit Books, 2014. The main character travels back and forward in time, always creating alternate iterations of himself. He eventually falls in love with himself, and mayhem ensues. (Chapters 2 and 5.)

William Gibson. *Neuromancer*. New York: Ace Books, 1984. Cyberspace is where all the cool posthumans hang out until the end of the universe. (Chapter 10.)

Joe Haldeman. *Forever War*. New York: St. Martin's Press, 1974. This novel is jam-packed with relativistic time travel (time dilation). Each battle in a war takes the hero centuries from the earth he remembers. (Chapter 1.)

Lucy Hawking and Stephen Hawking. *George and the Big Bang*. New York: Doubleday Children's Books, 2011. This middle-grade book examines the origins of the universe. (Chapter 4.)

Robert A. Heinlein. "The Man Who Sold the Moon." *The Man Who Sold the Moon*. Chicago: Shashta, 1950. Examples of nuclear and chemical rockets. (Chapter 17.)

Bruce T. Holmes. *Anvil of the Heart*. Evanston, IL: Haven Corp., June 1983. A novel about optimizing the genes of children, or else! (Chapter 9.)

Frank K. Kelly. "Star Ship Invincible." *Astounding Stories of Science Fiction* 9, no. 9 (January 1935). The word "starship" is first used. (Chapter 17.)

Nancy Kress. *After the Fall, Before the Fall, During the Fall*. San Francisco: Tachyon Publications, 2012. A novel about ecological disaster (chapter 11), dying before reaching adulthood, and time travel (chapter 1). What more is needed?

Nancy Kress. *Beggars in Spain*. New York: William Morrow, 1991. A story about a new genetic trait giving an edge to a small portion of humanity. (Chapter 9.)

Ann Leckie. *Ancillary Justice*. London: Orbit Books, 2013. Reversed transhumanism. A starship AI who uses humans as ancillary units becomes trapped in one of them. (Chapter 13.)

Ursula K. Le Guin. *Rocannon's World.* New York: Ace Books, 1966. The fictional ansible is introduced for instantaneous long-distance communications. (Chapter 16.)

Madeleine L'Engle. *A Wrinkle in Time.* New York: Farrar, Straus & Giroux, 1962. This young adult novel uses a tesseract to fold time for instantaneous travel across space. (Chapter 1.)

Jonathan Maberry. *Patient Zero.* New York: St. Martin's Griffin, 2009. A human-made disease zombifies humans. Not a smart choice. (Chapter 9.)

Anne McCaffrey. *The Ship Who Sang.* New York: Walker, 1969. Babies with birth defects have their brains wired into starships (transhumanism) to become the pilots. (Chapter 10.)

Seanan McGuire. "Each to Each." *Lightspeed Magazine: Women Destroy Science Fiction!* no. 49 (June 2014). Women who join the military are (politically) nudged into the navy, where they are transformed into mermaids (transhumanism) to be both pretty for the cameras (good for public relations) and efficient in ocean warfare. There is a promise to be converted back—only none have. (Chapter 9.)

China Miéville. *The City & the City.* London: Macmillan, 2009. Overlapping dimensions. (Chapter 3.)

Alan Moore. "Watchmen." DC Comics, no. 1–12 (September 1986–October 1987). Because sometimes you can see the end coming. Or not. (Chapter 21.)

Audrey Niffenegger. *The Time Traveler's Wife.* San Francisco: MacAdam/Cage, 2003. Time travel as a genetic disorder is mentioned. This plot idea is not hard science. (Chapter 1.)

Larry Niven. *Ringworld.* New York: Ballantine Books, 1970. A civilization so advanced that they build an inhabitable artificial ring around a sun. This sounds a lot like a Dyson sphere. (Chapter 6.)

Larry Niven and Dean Ellis. *All the Myriad Ways.* New York: Ballantine Books, 1971. The title story of this short story collection features the many-worlds interpretation of quantum mechanics. (Chapters 2 and 5.)

Larry Niven and Jerry Pournelle. *Footfall.* New York: Del Rey Books,

1985. This includes a ship propelled by nuclear bombs. There is also alien life. (Chapters 15 and 17.)

Claire North. *The First Fifteen Lives of Harry August.* Boston: Little, Brown, 2014. Time travel blended with the many-worlds interpretation of quantum physics. (Chapters 2 and 5.)

Ada Palmer. *Too Like the Lightning.* New York: Tor Books, 2016. Sometimes it is all about who controls the air highways. (Chapter 17.)

Hannu Rajaniemi. *The Quantum Thief* (Part One of the Jean le Flambeur trilogy). New York: Tor Books, 2011. It includes quantum computing (chapter 2) and artificial intelligence (chapter 13). There are posthumans hanging around who are radically different from baseline humans and from each other (chapter 10).

Kim Stanley Robinson. *Red Mars, Green Mars,* and *Blue Mars* (the Mars trilogy). New York: Spectra/Bantam Dell/Random House, 1993, 1994, and 1996. This is how you write about terraforming. (Chapter 12.)

Carl Sagan. *Contact.* New York: Simon and Schuster, 1985. First contact between humans and an extraterrestrial, an intelligent communicating race gives rise to moral/political issues. (Chapters 15 and 16.)

Hiroshi Sakurazaka. *All You Need Is Kill.* San Francisco: Haikasoru/VIZ Media, 2011. A soldier is trapped in a time loop, making changes to the timeline with each redo. Loosely the basis for the movie *Edge of Tomorrow.* (Chapter 5.)

Robert J. Sawyer. *Hominids* (the first book of the Neanderthal Parallax series). New York: Tor Books, 2003. Scientists discover a parallel Earth (chapter 5) where Neanderthals are the last of the homo genus. It is the Homo sapiens who have gone extinct. (A twist on chapter 8.)

Robert J. Sawyer. *Rollback.* New York: Tor Books, 2007. Genetic age reversal (chapter 9) and the reason to stay so young: long delayed communication with an alien race. (Chapter 16.)

John Scalzi. *Lock In.* New York: Tor Books, 2014. This book includes viruses, virtual reality, and posthumanism. (Chapters 8, 9, and 10.)

Zvi Schreiber. *Fizz: Nothing Is as It Seems.* Bala Cynwyd, PA: Zedess Publishing, 2011. In this time-travel novel, a girl meets famous scientists. It has many examples of the science from various chapters.

Garrett P. Serviss. *Edison's Conquest of Mars*. Los Angeles: Carcosa House, 1947 (New York Journal serialization 1898). This story includes asteroid mining. (Chapter 12.)

Mary Shelley. *Frankenstein; or, The Modern Prometheus*. London: Lackington, Hughes, Harding, Mavor, & Jones, 1818. This is about hacking life, nineteenth-century style. (Chapter 9.)

Masamune Shirow. *Ghost in the Shell*. Milwaukie: Dark Horse Comics, 2004. Japanese manga about identity and the interface between the human brain and technology. (Chapter 10.)

Dan Simmons. *Ilium* and *Olympos*. New York: HarperTorch, 2005; Harper Voyager, 2006. In these books posthumans recreate historic events. The story takes place on the terraformed planet Mars. (Chapters 8 and 12.)

Olaf Stapledon. *Last and First Men*. London: Methuen, 1930. Records a fictional history of humanity. (Chapter 21.)

Olaf Stapledon. *Star Maker*. London: Methuen, 1937. The first instance of what would later be called the Dyson sphere (chapter 6). There is a description of the multiverse (chapter 5).

Neal Stephenson. *Cryptonomicon*. New York: Avon, 1999. Computers and cryptography. You want more? Fictionalized Alan Turing and Albert Einstein drop in. (Chapter 13.)

Charles Stross. *The Family Trade*. (The first book of the Merchant Princes series.) New York: Tor Books, 2004. The series is about a drug-dealing family with an inheritable trait that grants the ability to travel between parallel Earths. (Chapter 5.)

Charles Stross. *Glasshouse*. New York: Ace Books, 2006. A twenty-seventh-century man agrees to an experiment. He wakes up in what appears to be the middle twentieth century in the body of a woman. There is also an interesting storyline about gender discrimination in the 1950s. (Chapter 9.)

Jules Verne. *From the Earth to the Moon*. Paris: Pierre-Jules Hetzel, 1865. Fictional use of light as a means of propulsion. (Chapter 17.)

Vernor Vinge. *True Names*. New York: Dell, 1981. This is an early example of a fully described cyberspace. It includes examples of the technological singularity. (Chapter 13.)

Kurt Vonnegut. *Galápagos*. New York: Dell, 1985. A question on the merits of the human intelligence in evolution. In the distant

future, humans have evolved into sea creatures that laugh at farts. (Chapter 8.)
Andy Weir. *The Martian*. Self-published, 2011; New York: Crown, 2014. This is an example of surviving in a hostile environment relying on limited terraforming. (Chapter 12.)
H. G. Wells. *The First Men in the Moon*. Oxford: Oxford University Press, 2017. First published in *Strand Magazine* from December 1900 to August 1901. Space travel using a fictional material called cavorite that negates gravity. (Chapters 17 and 18.)
H. G. Wells. *The Time Machine*. London: William Heinemann, 1895. The Morlocks and the Eloi as our potential posthuman future. (Chapter 9.)
H. G. Wells. *War of the Worlds*. London: William Heinemann, 1898. Unplanned by humans, it is the common cold (a biological weapon) that saves the earth for humans. (Chapter 9.)
Kate Wilhelm. *Where Late the Sweet Birds Sang*. New York: Harper, 1976. Environmental changes and disease lead to biological manipulations and cloning. (Chapters 8, 9, and 11.)
Jack Williamson. "Collision Orbit." *Astounding Science Fiction* 29, no. 5 (July 6, 1942). This short story contains the first use of the term "terraforming." (Chapter 12.)

MOVIE LIST

10,000 BC. Directed by Roland Emmerich. Warner Bros., 2008. (Chapter 8.)
2001: A Space Odyssey. Directed by Stanley Kubrick. Metro-Goldwyn-Mayer, 1968. (Chapter 15.)
A.I. Artificial Intelligence. Directed by Steven Spielberg. Warner Bros., 2001. (Chapter 13.)
Alien. Directed by Ridley Scott. 20th Century Fox, 1979. (Chapter 15.)
All You Need Is Kill (see reading list, chapter 5).
An Inconvenient Truth. Directed by Davis Guggenheim. Paramount Classics, 2006. (Chapter 11.)
The Andromeda Strain. Directed by Robert Wise. Universal, 1971. (Chapter 9.)

Another Earth. Directed by Mike Cahill. Fox Searchlight, 2011. (Chapter 2.)

Apollo 13. Directed by Ron Howard. Universal, 1995. (Chapter 17.)

Arrival. Directed by Denis Villeneuve. Paramount Pictures, 2016. (Chapter 15.)

Blade Runner. Directed by Ridley Scott. Warner Bros., 1982. (Chapter 9.)

The Butterfly Effect. Directed by Eric Bress and J. Mackye Gruber. New Line Cinema, 2004. (Chapter 5.)

Cargo. Directed by Ivan Engler. Atlantis, 2009. (Chapters 11 and 20.)

Close Encounters of the Third Kind. Directed by Steven Spielberg. Columbia, 1977. (Chapter 15.)

Coherence. Directed by James Ward Byrkit. Oscilloscope Laboratories, 2013. (Chapter 2.)

Contact. Directed by Robert Zemeckis. Warner Bros., 1997. (Chapter 15.)

The Day After Tomorrow. Directed by Roland Emmerich. 20th Century Fox, 2004. (Chapter 11.)

The Day the Earth Stood Still. Directed by Robert Wise. 20th Century Fox, 1951. (Chapter 15.)

Deep Impact. Directed by Mimi Leder. Paramount Pictures, 1998. (Chapter 21.)

Edge of Tomorrow. Directed by Doug Liman. Warner Bros., 2014.

Elysium. Directed by Neill Blomkamp. TriStar, 2013. (Chapter 11.)

Eternal Sunshine of the Spotless Mind. Directed by Michael Gondry. Focus, 2004. (Chapter 20.)

Europa Report. Directed by Sebastian Cordero. Magnet Releasing/Magnolia Pictures, 2013. (Chapter 15.)

Ex Machina. Directed by Alex Garland. Universal, 2015. (Chapter 13.)

Fantastic Voyage. Directed by Richard Fleischer. 20th Century Fox, 1966. (Chapter 10.)

Forbidden Planet. Directed by Fred M. Wilcox. Metro-Goldwyn-Mayer, 1956. (Chapter 14.)

Gattaca. Directed by Andrew Niccol. Columbia, 1997. (Chapter 9.)

Gravity. Directed by Alfonso Cuaron. Warner Bros., 2013. (Chapter 17.)

Hidden Figures. Directed by Theodore Melfi. 20th Century Fox, 2016. (Chapter 13.)

Inception. Directed by Christopher Nolan. Warner Bros., 2010. (Chapter 20.)

Interstellar. Directed by Christopher Nolan. Okotoks, Alberta, Canada: Paramount Pictures, 2014. (Chapter 1.)

Iron Man. Directed by Jon Favreau. Paramount Pictures, 2008. (Chapter 10.)

The Lawnmower Man. Directed by Brett Leonard. New Line Cinema, 1992. (Chapter 20.)

The Man from Earth. Directed by Richard Schenkman. Anchor Bay Entertainment/Shoreline Entertainment, 2007. (Chapter 8.)

The Matrix. Directed by the Wachowski Brothers. Warner Bros., 1999. (Chapter 20.)

Metropolis. Directed by Fritz Lang. UFA, 1927. (Chapter 13.)

Moon. Directed by Duncan Jones. Sony, 2009. (Chapter 9.)

Mr. Nobody. Directed by Jaco Van Dormael. Wild Bunch, 2009. (Chapter 3).

Oblivion. Directed by Joseph Kosinski. Universal, 2013. (Chapter 9.)

One Million Years BC (1966 version). Directed by Don Chaffey. 20th Century Fox, 1966. (Chapter 8.)

The Perfect Storm. Directed by Wolfgang Petersen. Warner Bros., 2000. (Chapter 11.)

Planet of the Apes. Directed by Franklin J. Schaffner. 20th Century Fox, 1968. (Chapter 8.)

Robocop. Directed by Paul Verhoeven. Orion, 1987. (Chapter 10.)

Robot & Frank. Directed by Jake Schreier. Samuel Goldwyn Films, 2012. (Chapter 14.)

Silent Running. Directed by Douglas Trumbull. Universal, 1972. (Chapter 11.)

Solaris. Directed by Andrei Tarkovsky. Soviet Union, 1972.

Star Trek IV: The Voyage Home. Directed by Leonard Nimoy. Paramount Pictures, 1986. (Chapter 18.)

Star Wars. Directed by George Lucas. 20th Century Fox, 1977. (Chapter 10.)

The Terminator. Directed by James Cameron. Orion, 1984. (Chapter 14.)

Total Recall. Directed by Paul Verhoeven. TriStar, 1990. (Chapter 20.)
Twelve Monkeys. Directed by Terry Gilliam. Universal, 1995. (Chapter 9.)
Twister. Directed by Jan de Bont. Warner Bros., 1996. (Chapter 11.)
Woman in the Moon. Directed by Fritz Lang. UFA, 1929. (Chapter 17.)

SONG LIST

The B-52's, "There's a Moon in the Sky (Called the Moon)," written by the B-52's, released July 6, 1979, on *The B-52's*, Warner Bros.
Barenaked Ladies, "History of Everything," written by Ed Robertson, Jim Creeggan, Kevin Hearn, and Tyler Stewart, released October 9, 2007, theme for television series *The Big Bang Theory*.
Blackalicious (featuring Cut Chemist), "Chemical Calisthenics," written by Lucas Christian MacFadden, Timothy Jerome Parker, on *Blazing Arrow*, released April 20, 2002, Upper Cut Music.
David Bowie, "Space Oddity," written by David Bowie, released July 11, 1969, on *David Bowie*, Phillips.
David Bowie, "Starman," written by David Bowie, released April 28, 1972, on *The Rise and Fall of Ziggy Stardust and the Spiders from Mars*, RCA.
Broadside Electric, "Ampere's Law," written by Walter Smith, released May 14, 2002, on *Live: Do Not Immerse*, Walter Fox Smith.
The Buggles, "Living in the Plastic Age," written by Trevor Horn, Geoff Downes, released on January 14, 1980, on *The Age of Plastic*, Island.
Kate Bush, "Pi," written by Kate Bush, released November 7, 2005, on *Aerial*, Columbia (US).
Jarvis Cocker, "Quantum Theory," written by Jarvis Cocker, released April 3, 2007, on *Jarvis*, (US) Rough Trade Records.
Coldplay, "The Scientist," written by Guy Berryman, Jonny Buckland, Will Champion, and Chris Martin, released November 4, 2002, on *A Rush of Blood to the Head*, Capital (US).
Alice Cooper, "Clones (We're All)," written by David Carron, released May 1980, *Flush the Fashion*, Warner Bros.
Elvis Costello, "My Science Fiction Twin," written by Declan MacManus, released March 8, 1994, on *Brutal Youth*, Warner Bros.

Cracker, "Show Me How This Thing Works," written by David Lowery and Johnny Hickman, released on May 5, 2009, on *Sunrise in the Land of Milk and Honey*, 429 Records.

Thomas Dolby, "She Blinded Me with Science," written by Thomas Dolby and Jo Kerr, released 1982, on *The Golden Age of Wireless*, Capital.

Thomas Dolby, "Windpower," written by Thomas Dolby, released on March 1982, on *The Golden Age of Wireless*, Venice in Peril/EMI.

John Grant, "Outer Space," written by John Grant, released April 19, 2010, on *Queen of Denmark*, Bella Union.

Imagine Dragons, "Radioactive," written by Alexander Grant, Ben McKee, Josh Mosser, Daniel Platzman, Dan Reynolds, and Wayne Sermon, released October 29, 2012, on *Night Visions*, Interscope (US).

Elton John, "Rocket Man," written by Elton John, Bernie Taupin, released April 14, 1972, on *Honky Chateau*, Uni Records (US).

Tom Lehrer, "The Elements," written by Tom Lehrer, released 1959, on *More of Tom Lehrer*, Lehrer Records.

Kate and Anna McGarrigle, "NaCl," written by Kate McGarrigle, released 1978, on *Pronto Monto*, Warner Bros.

Oingo Boingo, "Weird Science," written by Danny Elfman, released October 28, 1985, on *Dead Man's Party*, MCA.

Katy Perry, "E.T.," written by Katy Perry, Lukasz Gottwald, Max Martin, Joshua Coleman, and Kanye West, released February 16, 2011, on *Teenage Dream*, Capitol.

The Polecats, "Make a Circuit with Me," written by Tim Worman and Phil Bloomberg, released 1983 on *Polecats Are Go!* Mercury.

Monty Python, "Galaxy Song," written by Eric Idle and John Du Prez, released on June 27, 1983, on *Monty Python's the Meaning of Life*, CBS/MCA.

Queen, "39," written by Brian May, released November 21, 1975, on *A Night at the Opera*, EMI/Elektra.

Radiohead, "Subterranean Homesick Alien," written by Thom Yorke, Jonny Greenwood, Ed O'Brien, Colin Greenwood, and Phil Selway, released on June 16, 1997, on *OK Computer*, Capitol.

R.E.M., "It's the End of the World as We Know It (And I Feel Fine)," written by Bill Berry, Peter Buck, Mike Mills, and Michael Stipe, released November 16, 1987, on *Document*, I.R.S.

The Rolling Stones, "2000 Light Years from Home," written by Mick Jagger and Keith Richards, released December 23, 1967, on *Their Satanic Majesties Request*, London/ABKCO.

Shriekback, "Recessive Jean," written by Barry Andrews, Carl Marsh, and Martyn Barker, released March 4, 2015, on *Without Real String or Fish*, Shriekback self-released.

Sia, "Academia," written by Sia Furler and Dan Carey, released January 8, 2008, on *Some People Have Real Problems*. Hear Music/Monkey Puzzle.

Soundgarden, "Black Hole Sun," written by Chris Cornell, released May 4, 1994, on *Superunknown*, A&M.

Styx, "Mr. Roboto," written by Dennis DeYoung, released February 11, 1983, on *Snowblind*, A&M.

They Might Be Giants, "My Brother the Ape," written by They Might Be Giants, released September 1, 2009, on *Here Comes Science*, Disneysound/Idlewild.

They Might Be Giants, "Science Is Real," written by They Might Be Giants, released September 1, 2009, on *Here Comes Science*, Disneysound/Idlewild.

INDEX

absolute zero, 241, 280, 282
Adams, Douglas, 119, 249, 267
adaptation, 110, 226
AI. *See* artificial intelligence
albedo, 153, 164
Aldiss, Brian, 179
Aldrin, Buzz, 205
Allen, Woody, 127
Alpha Centauri system, 166
Alvelda, Phillip, 254
Amara, Roy, 249
Anderson, Carl D., 57, 289
Andromeda galaxy, 79, 93, 199, 290, 307
anthropic principle
    strong anthropic principle, 87–88
    weak anthropic principle, 87
anthropic reasoning, 88
antimatter engine, 24, 56–57, 62, 69, 223–24, 230, 289, 303
antineutrino, 215
Arctic Rising (Buckell), 153, 297
Aristotle, 237
artificial intelligence (AI), 24, 64, 171, 176–81, 185–86, 188–89, 194–95, 221, 254, 262, 270, 283, 299
Artsutanov, Yuri N., 219
Asimov, Isaac, 20, 175, 186, 189, 194, 299–300
    introduced word "robotics," 186
    See also Clarke-Asimov treaty; Three Laws of Robotics
astronomical unit (AU), 79–80, 171
atom, oldest, 56
atomic theory, 53, 55, 57, 59, 289
augmented reality, 253, 259
Australopithecus afarensis, 112

bacteriophages, 132
Barjavel, René, 49, 288

Barrow scale, 97
Bates, Harry, 197, 301
BECs. *See* Bose-Einstein Condensates
Beggars in Spain (Kress), 127, 294
beta decay, 215
big bang, 24, 28, 36, 56, 71–77, 80–81, 84, 86, 88, 234, 240, 279, 290
black holes, 24, 33, 36–37, 64, 76–77, 96, 99–108, 202, 211, 266–67, 279, 282, 287–89, 293
    evaporating, 65, 104, 289
    spinning, 106
Bohr, Niels, 46
Bose-Einstein Condensates (BECs), 236, 304
branes, 64, 74, 80, 290. *See* membrane theory
Briggs, Gordon, 190, 301
Burroughs, William S., 135
Byron, Lord, 19, 176

Caesar, Sid, 255
Calabi-Yau space, 64
Cambrian Explosion, 116, 271, 294
Camden, Rodger, 127
Capek, Karel, 186
    introduced word "robot" in science fiction, 186
carbon sequester, 151, 163
Card, Orson Scott, 27, 29, 210, 287
Casimir, Hendrik, 95
Casimir effect, 95
Cepheid stars, 70, 82
cerebellum ages slower than other parts of body, 128
CFCs. *See* chlorofluorocarbons
Chandrasekhar limit, 100
Chaotic Inflation, 88
chlorofluorocarbons (CFCs), 156

civilizations, 24, 91, 93–95, 97, 140, 197, 199–201, 205, 209, 214, 279, 292
Cixin, Liu, 64, 289, 297
Clairmont, Claire, 19
Clarke, Arthur C., 20, 25, 138, 175, 197, 219, 223, 283, 287, 295, 303, 308. *See also* Clarke-Asimov treaty
Clarke, Roger, 189
Clarke-Asimov treaty, 20
cloning, 130
Clustered Regularly Interspaced Short Palindromic Repeats (CRISPR), 126, 294
CMB. *See* cosmic microwave background
cognition, 174, 181, 192
collapsing of the wave function, 43
companion star to the sun, 273
computer identification, 254
consciousness, 124, 141, 171–74, 178, 180–81, 258
conservation of energy, 280–81
Cooper, Joseph, 32
Copenhagen interpretation of quantum mechanics, 43–45, 48
cosmic inflation, 75, 77, 88
cosmic microwave background (CMB), 72, 75–77, 290
cosmic strings, 36–37, 288
cosmology, 24, 67, 69, 80, 86
crime-busting science, 133
CRISPR. *See* Clustered Regularly Interspaced Short Palindromic Repeats
cyanobacteria, 116
cybering, 141
cyberspace, 141

Daniels, Paige, 34, 288
dark matter, 75–76, 205, 233–34, 274
DARPA. *See* Defense Advanced Research Projects Agency
Darwin, Charles, 109
DC Comics universe, 83, 93, 137
decoherence, 44–45
Defense Advanced Research Projects Agency (DARPA), 254, 306
Delany, Samuel R., 226
Descartes, René, 171, 180, 185

devil facial tumor disease (DFTD), 111
DFTD. *See* devil facial tumor disease
Dick, Philip K., 21, 130, 257, 263
Dirac, Paul, 57
DNA (deoxyribonucleic acid), 109–21, 123, 125–29, 132–34, 140, 156, 168, 294–95
    sequences, 125, 133
Doctor Who (TV series), 25, 27, 35, 61, 93, 136–37, 140, 260, 276, 286, 289, 296, 306
Drake, Frank, 199
Drake equation, 199–201
dwarf planet, 79–80
Dylan, Bob, 125
Dyson, Freeman, 92, 292
Dyson sphere, 92, 93

Earth extinctions
    Cretaceous-Paleogene, 271, 275
    Holocene extinction, 272
    Late Devonian, 271
    Ordovician-Silurian, 271
    Permian-Triassic, 271
    Triassic-Jurassic, 271
Eddington valve, 82
Einstein, Albert, 23, 27, 34, 287
Einstein-Rosen bridges. *See* wormholes
electromagnetic drive, 224, 230
electromagnetic force, 53, 55–56, 58
electromagnetic spectrum, 71
electromagnetism, 54, 59, 107, 232–33
emergent property, 172–73, 176
Ender's Game (Card), 27, 287
energy-mass equivalency, 24, 33
energy returned on energy invested (EROEI), 97–98
entropy, 270, 280–82
epigenetic traits, 121
erasing memories, 264
EROEI. *See* energy returned on energy invested
escape velocity, 100, 102, 302
eukaryote cell, 116, 120
event horizon, 103–104, 107, 267
Everett, Hugh, 45, 288

evolution theory, 109
exoplanet, 165–66

faster-than-light (FTL), 17, 47, 213
Fermi, Enrico, 200
Fermi Paradox, 200
Feynman, Richard, 128
Fountains of Paradise, The (Clarke), 219, 303
Fourier, Joseph, 149
four universal forces
    electromagnetic force, 53, 55–56, 58
    gravity, 23–24, 30, 32–35, 37, 54, 59, 61–66, 75–76, 80–82, 87, 95, 99, 100, 102–103, 106–107, 109, 159, 160, 163–64, 175, 203, 210, 211, 214, 217, 219, 219, 225–28, 231–35, 266, 273–74, 276–79
    strong nuclear force, 54, 56, 58–59, 87, 215, 234
    weak nuclear force, 54, 59, 214, 234
frame-dragging, 106–107
Frankenstein (Shelley), 19, 125, 146
Frankenstein, Victor, 125
From the Earth to the Moon (Verne), 19
FTL. *See* faster-than-light

galactic distances, 70
Galapagos (Vonnegut), 110
gene-editing technology, 125
gene manipulation, 127, 133
gene modification, 126
general relativity, 23, 30–33, 36–37, 64, 69, 72, 76–77, 99, 103–108, 159, 210, 226, 246, 267, 281
genetic engineering, 151–52, 154
geoengineering, 151–52, 154
geologic timeline/epochs, 155
Gerrold, David, 85, 291
Gibson, William, 138, 141
    credited with word "cyberspace," 141
Gionfriddo, Kaiba, 252
global warming, 24, 150–54, 175, 179, 297
Godwin, Mary Wollstonecraft, 19
Goldilocks zone, 159, 165–66
gravitational lensing, 105

gravitational waves, 36, 76–77, 211, 279, 290
gravitons, 54, 214
greenhouse effect, 81, 147–48, 155, 163, 277, 297
grid cells, 174
Guth, Alan, 75

hadron, 58, 74, 233
Haldeman, Joe, 29, 287
Hawking, Stephen, 36, 74, 99, 103, 228, 258, 270, 290, 303, 306–307
Hawking radiation, 103
Hayflick, Leonard, 129
Hayflick limit, 129, 270
heat death, 282
Heinlein, Robert, 20
Heisenberg, Werner, 41–42
Heisenberg uncertainty principle, 41, 48, 94
Higgs field, 232–33
Hitchhiker's Guide to the Galaxy, The (Adams), 119, 177, 195
holographic principle, 265–67
holographic reality, 265
hominid, 111, 270
hominin, 111–12, 114, 120, 125, 197
Homo floresiensis, 113
Homo heidelbergensis, 112
Homo sapiens, 83, 86, 113, 120
Hoyle, Fred, 71
Hubble, Edwin, 71, 176, 290
Hubble's constant, 67
human story, 111
Hynek, Josef Allen, 206, 302

IEDs. *See* improvised explosive devices
Illium (Simmons), 140, 223
Imperial Earth (Clarke), 20
improvised explosive devices (IEDs), 187
Industrial Revolution, 18, 272
information paradox, 108, 267
infrared (IR) light, 147
Integrated Tissue and Organ Printing (ITOP), 251
International Space Station (ISS), 225, 232

334 INDEX

ion engine, 209, 223, 230
ISS. *See* International Space Station
ITOP. *See* Integrated Tissue and Organ Printing

Jaynes, Julian, 180, 299
Journey in Other Worlds, A (Astor), 217, 302

Kardashev, Nikolai, 92
Kardashev scale, 96
kinetic energy, 74, 235–36, 241, 275, 280
Kovacs, Walter, 269
Kubrick, Stanley, 179

Lamoreaux, Steve K., 95
Large Hadron Collider (LHC), 74, 233
Laser Interferometer Gravitational-Wave Observatory (LIGO), 77, 290
Leavitt, Henrietta Swan, 70, 176, 290, 299
Le Guin, Ursula K., 21, 41, 61, 210
Lennon, John, 174, 258
leptons, 58, 97
Le Voyageur Imprudent (Barjavel), 49, 288
light-energy weapons, 250
LIGO. *See* Laser Interferometer Gravitational-Wave Observatory
Little Earnest Book upon a Great Old Subject, A (Wilson), 19, 287
Local Group, 79, 278–79
Lock In (Scalzi), 141, 296
loop quantum gravity (LQG) theory, 24, 64–66, 289
Lovelace Test, 178

Maathai, Wangari, 178
macrophage, 132
magnetism, 160
Malthus, Thomas Robert, 110, 293
many-worlds interpretation, 45–46, 48–50, 85
material engineering, 237
Marvel Comics, 83, 93, 137
Matisse, Henri, 175
membrane dimensions, 64
membrane theory, 64, 74, 86. *See* branes

metamaterial, 237, 245–47
metamorphosis, 134
microgravity, 160
Milky Way galaxy, 28, 93, 205, 278, 292
Milner, Yuri, 228, 303
Moore, Gordon, 182
Moore's law, 181–82, 300
Mount Tambora, eruption of, 19
M-theory, 63–64, 66
mutants, 39

nanite, 139–40
nanotechnology, 97, 123, 132, 136, 139, 243, 245
Nasa's Gravity Probe B satellite, 106
National Centers for Environmental Information (NCEI), 149, 296
National Oceanic and Atmospheric Administration (NOAA), 149, 296–98
natural selection, 109–11, 115, 120, 125, 140
NCEI. *See* National Centers for Environmental Information
near-Earth objects (NEOs), 275
negative energy, 96, 104
negative greenhouse effect, 155, 297
NESD. *See* Neural Engineering System Design
Neural Engineering System Design (NESD), 254–55, 306
Neuromancer (Gibson), 141
neurons, 138, 172–75, 186, 255, 264
neutrino, 24, 214–15, 233–34
neutrino beam communication, 214
Niccol, Andrew, 127
Nietzsche, Friedrich, 123
Niven, Larry, 17, 92, 219
NOAA. *See* National Oceanic and Atmospheric Administration
nonillion, 78, 282
nuclear fusion, 59, 73, 81, 273
nuclear propulsion, 222, 230

observable universe, 63, 67, 69, 71–72, 93, 199
Obst, Lynda, 32

Olbers, Heinrich Wilhelm, 81
Olbers's Paradox, 81
Olympos (Simmons), 140, 223
Ophiocordyceps (fungus spore), 131
optical-mining, 162
Orgel, Leslie, 109
ozone layer, 156

parallel worlds, 24, 83–85, 87–89, 291
parsec, 67–68
period-luminosity relationship, 70
PET. *See* positron-emission tomography
photosynthesis, 50, 115–16, 159, 166, 206
Planck, Max, 63
Planck length, 66
Podkletnov, Eugene, 228, 303
Polidori, John William, 19
positron-emission tomography (PET), 57
posthumanism, 124, 140–42
Practice Effect, The (Brin), 86, 291
Proxima Centauri, 166–67

quantum communication, 46–47, 49, 211–13
quantum computers, 47–49, 214
quantum computing, 24, 46–48, 182
quantum encryption, 213
quantum entanglement, 24, 46–47, 80, 105, 212–14, 288
quantum immortality, 50–51
quantum leap, 44
quantum particles, 212
quantum physics, 43, 45, 48, 213
quantum suicide, 24, 50–51
quantum teleportation, 24, 212
quantum trivia, 48
quark, 57–58, 62, 75, 97, 232–33
qubits, 47

ramjet, 218
Rare Earth Hypothesis, 202, 207
redshift, 71
Rees, Martin, 271, 307
Report on Planet Three and Other Speculations (Clarke), 20, 287
Rex, Clyde, 134

Ringworld (Niven), 92
RNA (ribonucleic acid), 118–19, 121, 126
Robinson, Kim Stanley, 17, 165
robocops, 193
robotic evolution, 187
robotic tele-surgery, 192
Rodgers, Aaron, 102

Sagan, Carl, 67, 145, 164
Sagittarius A*, 28, 78, 99
Sakurazaka, Hiroshi, 83
Sartre, Jean-Paul, 173
Scalzi, John, 141, 296
Scheutz, Matthias, 190, 301
Schrödinger cat thought experiment, 50
Schrödinger wave equation, 42, 48, 85
Schwarzschild, Karl, 102
SCNT (somatic cell nuclear transfer) method, 130
scramjet, 218
Seldon, Hari, 175
Shaw, George Bernard, 209
Shay, Susanne, 49, 286
Shelley, Mary, 19, 21, 125
Shelley, Percy Bysshe, 19
Shirow, Masamune, 137
shrinking horizon, 70
singularity, 33, 36, 71, 100–102, 105, 107
Snicket, Lemony, 167
solar sails, 19, 221–22, 229–30, 303
solar system, 68, 79–81, 159, 163, 165–66, 199, 202, 211, 228, 273, 278, 291, 301, 307
solar wind, 160, 163–64, 221
Solo, Han, 68
space elevator, 20, 219–21, 229
spacetime, 27, 29, 31–37, 62, 64, 66, 69, 71, 74–77, 80, 99–100, 105–107, 109, 210–11, 214, 225, 234, 282
spaghettification, 103, 107
spatial cloaking, 244–45, 247
spatial entanglement, 51
special relativity, 27–30, 32–33, 35, 37–38, 96, 175, 209, 226
special relativity paradox, 37
speciation, 119

standard candles (cosmic yardsticks), 70
Stapledon, Olaf, 92, 292
Star Maker (Stapledon), 92, 292
Star Trek universe, 69, 119, 140, 210, 224, 250–51
steady state, 71
stem cell, 129, 251
STEM experts, 17
string theory, 23–24, 61–66, 74, 80, 84, 86–88, 210, 214
Stross, Charles, 83, 140, 157, 291, 296
Sturgeon, Theodore, 28
subatomic uncertainty, 51
sun, history of our, 81
superatom, 236, 240–41
superconductor, 228, 241–43, 304
super-galaxy, 279
supernova, 56, 70, 76, 100
superposition, 43–45, 47, 49–50, 213
supersized galaxies, 278
survival of the fittest, 110

tachyon particles, 96
technological singularity, 178–81, 299
tectonic plates, 161
temporal cloaking, 244–47
terraforming, 24, 68, 158, 163–65, 167–68, 278, 298
thermodynamics, 81, 239, 280–82
Thiel, Peter, 249
Thorne, Kip, 32
Three-Body Problem, The (Cixin), 64, 154, 289, 297
three-dimensional cube, 24
three-dimensional universe, 62, 64, 74, 265
3-D printing, 251–52
Three Laws of Robotics, 189
time dilation, 27, 29–32, 69, 107–108
time-traveling text messages, 24, 51
time-travel paradox, 49
transhuman brain, 138
transhumanism, 124, 141–42, 152, 157, 226
transit method, 166, 315
turbojet, 218

twentieth-century physics
  Einstein's theories, 23
  quantum mechanics, 23–24, 41–50, 54, 57, 61–66, 73, 80, 85, 94–95, 105–109, 210, 214, 236, 258, 267, 285, 288
Twenty Thousand Leagues under the Sea (Verne), 19
two-dimensional string, 36
two-quark combination, 58
Tyndall, John, 149

ufology, 206
ultraviolet (UV) light, 147
unidentified flying objects (UFOs), 206
universal Dark Age, 76
universe, facts about our, 77

Vader, Darth, 137, 233
Verne, Jules, 19
virtual particles, 24, 94–96, 103, 224, 292
volatiles, 162, 165, 168
volcanic winter, 19, 113, 146

Water Knife, The (Bacigalupi), 154, 297
Watts, Peter, 207
wave-particle duality, 42, 52, 56, 212
weakly interacting massive particles (WIMPs), 234
Wheeler, John, 31
Williamson, Jack, 159, 298
WIMPs. *See* weakly interacting massive particles
wobble method, 166
wormholes, 34

zero-point energy (vacuum energy), 24, 94–96
zeta-pseudosubstrate inhibitory peptide (ZIP), 264
ZIP. *See* zeta-pseudosubstrate inhibitory peptide
zombie apocalypse, 131